High Tech, High Hope

Upside Books examine events in business and management through the lens of technology. *Upside Magazine* is the preeminent magazine for executives and managers eager to understand the business of high tech.

Published:

High Tech, High Hope: Turning Your Vision of Technology into Business Success, Paul Franson

Risky Business: Protect Your Business from Being Stalked, Conned, Libeled, or Blackmailed on the Web, Daniel Janal

Forthcoming:

Silicon Gold Rush: The Next Generation of High-Tech Stars Rewrite the Rules of Business, Karen Southwick

PEOPLE · TECHNOLOGY · CAPITAL

UPSIDE
WWW.UPSIDE.COM

High Tech, High Hope

TURNING YOUR VISION OF TECHNOLOGY INTO BUSINESS SUCCESS

PAUL FRANSON

John Wiley & Sons, Inc.

New York ● Chichester ● Weinheim ● Brisbane ● Singapore ● Toronto

Copyright © 1998 by Paul Franson. All rights reserved.
Published by John Wiley & Sons, Inc.

Published simultaneously in Canada.

This publication is designed to provide accurate and authoritative information in regard to the subject matter covered. It is sold with the understanding that the publisher is not engaged in rendering legal, accounting, or other professional services. If legal advice or other expert assistance is required, the services of a competent professional person should be sought.

Designations used by companies to distinguish their products are often claimed as trademarks. In all instances where John Wiley & Sons, Inc. is aware of a claim, the product names appear in initial capital or all capital letters. Readers, however, should contact the appropriate companies for more complete information regarding trademarks and registration.

Library of Congress Cataloging-in-Publication Data:

Franson, Paul, 1941–
    High tech, high hope : turning your vision of technology
into business success / Paul Franson.
        p.   cm.
    Includes bibliographical references.
    ISBN 0-471-23981-X (alk. paper)
    1. Technological innovations.   I. Title.
HD45.F717   1998   97-31546
658.5'14—dc21   CIP

Printed in the United States of America.
10 9 8 7 6 5 4 3 2 1

Contents

Foreword

Before *Upside* magazine was launched in 1989, coverage of high technology meant lots and lots of information on product upgrades (version 2.1 of this or that), but little insight into what really made the industry tick. The kind of entrepreneur who was starting a technology company in a brand-new niche with hopes of creating a $1 billion enterprise was an obscure enigma. Companies and people who have become household words to a wired world—Microsoft and Bill Gates, Intel and Andy Grove—were known mostly to a then-small group of trade press and high-tech analysts. *Upside* changed all that by focusing on technology in a broader way—as a means of transforming business and society, which indeed it has. The magazine also introduced the people who were making technology as the heroes, or in some cases the antiheroes, of the coming twenty-first century. For technology companies were not only creating products that would enable, indeed force, profound changes in the way all businesses operate, they were creating a new corporate culture—egalitarian, fast-moving, risk-embracing, decentralized.

Now Upside Media Inc., the parent of *Upside* magazine, is bringing that same candid perspective to the world of business books. Under the Upside imprimatur, published by John Wiley & Sons, we will deliver books that speak to how technology could and should be used at corporations, give an insider's view of the rise of entrepreneurial companies, and focus on what technological change will mean to you as an executive, an employee, and a human being. We define technology broadly, encompassing computers and software but also telecommunications, video games, and the Internet—generally, the whole spectrum of innovation jump-started by the invention of the microprocessor less than 30 years ago.

The first book in the Upside series, *High Tech, High Hope: Turning Your Vision of Technology into Business Success,* is a clear, brisk, no-holds-barred look at technology's power to remake corporate processes. Writer Paul Franson, an experienced journalist and former high-tech public relations consultant, presents wide-ranging examples of the impact of technology on everything from product development to manufacturing to human resources. He points out that every company is becoming a technology company. By breaking down artificial walls between departments, technologies like desktop computers and E-mail are empowering teams who are "connected by purpose instead of position."

Franson effectively ties specific product developments such as client-server computing and networking to their place in the corporate evolutionary process, which has been vastly accelerated by the use of information technology. In accessible, but not simplistic, language, *High Tech, High Hope* details how corporations as diverse as Boeing, Wal-Mart, and EMI Records have harnessed technology to speed product design, improve inventory control, and boost distribution.

The title *High Tech, High Hope* actually embodies the vision of the Upside series. The spread of technology throughout the world, from cellular phones in China to Internet service providers in Brazil, has made Marshall McLuhan's global village even smaller. Will we use the Internet's power of instantaneous communication to understand each other better, or to spam someone with racially charged epithets? Is the technically endowed corporation moving closer to the mindless bureaucracy pictured so bitingly in Dilbert or to the ideals of Hewlett-Packard Company, whose "HP way" embodies respect for employees above all else?

There are many management theorists who pin their hopes on technology, ardently believing that it will make corporations more competitive, more productive, and more successful, while freeing employees from the drudgery of repetitive tasks that can be handled by computers. Then there are the neo-Luddites, who counter that technology applied indiscriminately only worsens existing problems in the business world, overlaying already-complex processes with a new burden of mindless routine and endless fixes.

The truth can be neither, or both. For technology has inescapably altered the evolution of the corporation, and in so doing, is being changed itself.

The Upside series, led by *High Tech, High Hope,* will tell you about technology's brilliant successes, but also about its flaws. We will tell you about the entrepreneurs, warts and all, who have created some of the most visible and important companies in the world. We will be honest about what technology can accomplish, and what it can't. We won't pretend that all these advances will suddenly thrust us into a techno-Eden, but neither will we dismiss the undeniable benefits of such innovations as E-mail, desktop computing, and integrated circuits. In short, we will follow the mission statement of our parent, Upside Media, which is: "to provide authoritative, insightful, provocative, and opinionated analysis of the business of technology. Through its magazine, Web site, books, and events, Upside is the essential resource for technology executives, entrepreneurs, investors, and savvy observers who want to cut through the hype and focus a critical eye on the most important business driving the world's economy."

David Bunnell, Publisher and CEO, Upside Media Inc.
Karen Southwick, Editor, Upside Books

Which Came First: Technology or the Corporation?

■ INTRODUCTION

There is hardly any aspect of corporate life that hasn't dramatically changed because of the computer and communications technology developed in the last few decades—but that's not news.

What's surprising is that only recently has information technology (IT) created fundamental changes in the way corporations operate. Many observers have talked about the changes corporations must make to meet competition and operate effectively in the era of rapid market and product changes, but there's been relatively little discussion of the way technology itself is changing corporations. Instead, writers and analysts have focused on the need for reorganizing and refocusing companies without properly emphasizing the way technology is initiating these changes and then forcing them to happen.

This book intends to bring the situation into focus. It discusses how corporations are really using and benefiting from modern technology, and how this technology is changing them to the core. In a nutshell, notes Louis Gerstner, chairman of IBM,[1] "Information technology effectively requires every institution in the world to become an information company as well as a product company. So whatever product you make and sell, you also have to make and sell information. Or at least, make and provide information."

■ HOW TECHNOLOGY HAS CHANGED CORPORATIONS

Forget the chicken and the egg. For corporate sociologists, the question is, "Which came first, the technology or the organization?"

Just as power technology caused the industrial revolution and in the process created vast corporations where once only cottages provided goods and services, today's information revolution is mutating corporations into new forms. Many are beginning to resemble the cottage collectives of yesterday, with empowered individuals and small teams united to serve common purposes, all tied together with modern technology. Today, modern corporations are abandoning cherished beliefs and venerable processes as new technologies and new ways of thinking make their old ways inefficient and irrelevant. This is also having a big impact on companies' finances. Hewlett-Packard has seen its sales, general, and administrative costs drop significantly due to its use of technology (see Figure I-1).

The computer revolution that began after World War II didn't really change the way corporations functioned. It changed many of the things they did, and it created many new products, some of which changed the jobs of individuals in corporations. But the traditional corporate organization continued much as it had for 150 years: product development, purchasing, manufacturing, warehousing, distribution, marketing, and sales, all supported by finance and

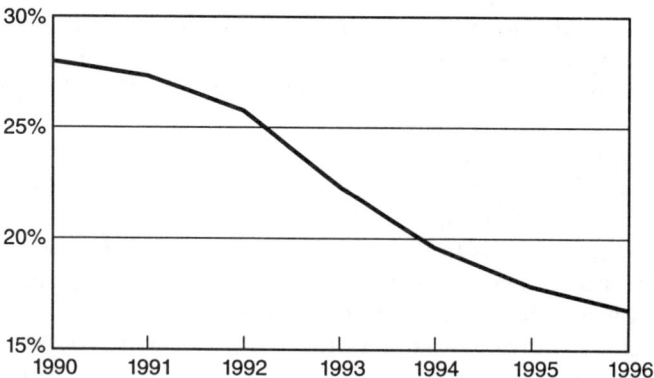

Figure I-1. Hewlett-Packard's sales, general and administrative costs have dropped significantly, partly from applying technology to its operations.

personnel. Each function remained independent, as the saying goes, "throwing things over the wall" to the next step in the corporate chain, little impacted by changes in the neighboring cells.

Technology changed that forever. And two interrelated technologies stand out as the critical factors that caused that change: networking and client/server computing. These technologies broke down the artificial walls in corporations, letting individuals and departments evolve into more natural teams—teams connected by purpose instead of position.

Says IBM's Gerstner, "The networked world holds the promise for the IT industry to focus on and deliver results for those things CEOs really worry about: speed, competitiveness, cycle time, market extension, globalization. This new network environment will move the IT industry off its preoccupation with making an individual clerical worker more productive and into the enterprise and solving enterprise problems."[2]

It's easy to underestimate the impact that networking has had on corporations, for that impact is far from complete. Many companies already have felt the effect: They've become more open, more efficient, and more effective, all the while becoming better places to work. Others haven't yet felt the shock, but they will—particularly as the most radical networking technology, the Internet, becomes pervasive.

Networking changes every person's value and job in the new corporate environment. Former low-level clerks and telephone operators become important customer contacts. High-paid middle managers find their positions superfluous. Assembly workers and truck drivers can save their companies millions of dollars. Executives can no longer hide their weaknesses behind precedence and privilege.

■ PROCESSES, NOT FUNCTIONS

Corporations everywhere are trying to find better ways to succeed in ever more competitive markets. As they undergo diagnosis and analysis from corporate doctors and analysts, they're realizing that a company's ways of doing things—its processes—are more important than its traditional functional organizations.

An example is the supply chain. More than the simple purchasing of yesterday, this concept starts with the customer (or the customer's customer), then proceeds to sales, procurement, manufacturing, warehousing, and distribution. When the customer orders, the impact is felt throughout the organization. It's not the pattern of the past, which started with purchasing products for manufacturing to build to inventory, with the hope that marketing and sales would generate sufficient—but not too many—orders.

Many companies don't even manufacture products in the old sense. They subcontract everything, so their "factory" is their purchasing department. There's a move to buy, not make. That's true not just for products sold, but also for services and even staff. Few corporations develop software for their own use. They buy it. They also hire janitorial and security firms, and increasingly are farming out critical functions such as benefit management, once considered a vital in-house service.

Microsoft's Bill Gates calls the integrated enterprise information a company's digital nervous system. "The whole boundary of what you do inside your company and what you do outside your company needs to be rethought once you have these digital nervous systems. I think the average size of a company will be a fair bit smaller than it is today. There'll be more things you decide can be outside the company than things you decide that are outside today that need to move in. And, of course, the boundaries won't be as hard as they have been. Because you'll be able to reach out to somebody who is far away and not only find them when they're far away, but also do work with them when they're far away, your flexibility will be much greater."

Gates says that on a trip to India he was stunned at the number of entrepreneurial companies there that see the Internet as a way for them to take the country's educated workforce and get world-level wages for those Ph.D.s and software developers. "In the past, the idea of going to India and doing software development was a very difficult thing. Would you find the right firm? Would the information flow back and forth very well? Now, with the Internet, you'll be able to go out there, see the firms that have skills, look at their references, see what people say about the work they did, see what their availability and their hourly rate is, their expertise, and conclude an arrangement to draw on that resource and get great software work done purely electronically."

He adds, "That kind of creativity, thinking through where the skills are outside your company anywhere around the world, can make a huge difference in terms of getting things done. It's sort of an extension of friction-free capitalism, taking the very mechanism of capitalism, which is the matching of buyers and sellers, and bringing that out onto the Internet."[3]

Product development is just as complex. In the past, a product sprang from a fertile imagination. A company made and marketed products with the hope that someone would buy them. Now most products evolve through an interactive process involving consumer research and careful attention to the competition. Designers create virtual products that exist only in computer memory. Their companies' sales representatives take those virtual products to customers for feedback and they take orders before the first physical model is built.

These sales reps have become part of the product development team, and they're ordering components for manufacturing when they take orders. They've also discovered the importance of others they never considered part of the sales team; the customer support clerk or technician in a company call center can make or break their million-dollar sales.

■ TECHNOLOGY IS STRATEGY

As traditional roles change, technology is growing in strategic importance to corporations. The wise ones are learning to exploit technology for their competitive advantage. Leading banks, airlines, retailing, and express services are examples of companies with superb information structures that may be more important to their overall success than marketing or sales, which were formerly considered the company's stars.

Other companies that have regarded technology only as a service or tool must recognize that it will become increasingly important to their success—even to their survival—as competitors raise the bar. No corporation will succeed without the intelligent use of technology, for it will be impossible to succeed at any fundamental strategy—be it low-cost supplier, innovator, customer champions, or niche master—without the clever use of information.

Louis Gerstner was a customer of information, a CEO in a consumer products company, before he joined IBM. "I believe that every CEO needs to worry: Is information technology—its power and threat—an integral part of your strategic thinking. Is it there at the table when you make strategic decisions? Do your senior managers understand it? Not bits and bytes, but the power to change? Is your CIO at the table when you're doing strategic planning? Are you on the leading edge? Are you going to be a follower? What are you going to do with your old distribution systems versus new distribution systems?"[4]

As corporations evolve from feudal kingdoms inhabited by fiefdoms with moats and walls around them, there's an increasingly greater need to coordinate the activities. Fortunately, modern technology makes this coordination possible and practical. Though few companies tie all of their functions together today, many are approaching that goal. The day is approaching when this once impossible dream will be commonplace.

The first applications of computers at most companies were for accounting, and that's where the computers hid for years. The traditional mainframe is great at processing numbers and data (but is expensive to buy and maintain), is difficult to program, and is best suited for one task at a time—even if it can be fooled into multitasking different activities. In this scenario, the mainframe was generally off-limits to other uses anyway. Other departments added computers for their own unique purposes, notably in engineering and technical applications, but they stayed out of mainstream business, limiting themselves to specialized applications.

■ THE IMPACT OF THE PERSONAL COMPUTER

The real change in corporate culture began with the personal computer. With the PC, any employee could have a computer of his or her very own and use it for real work. It simplified applications that were cumbersome with a mainframe, even without taking into account the problem of gaining access.

A mainframe required a skilled programmer to do things that a nontechnical user eventually could easily do with a spreadsheet on a PC. The forms and macros required to solve problems on PCs were trivial compared to traditional programming in COBOL or other computer languages. Soon PCs were ubiquitous among managers and professionals. PCs also moved into other arenas. They replaced dedicated word processors and started taking over database management functions that formerly required experts and "big" computers.

Management information system (MIS) managers reacted in horror as they saw these rogue computers serving important functions within their corporations. These PCs held vital information in inconsistent and inaccessible formats and were not secure from loss or damage.

Eventually, MIS departments connected PCs to their corporate mainframes, but primarily as replacements for dumb terminals. Some users, however, were more creative. They downloaded information from the mainframe so that they could manipulate it for their own uses—if not for anyone else's. PCs also became a conduit for electronic mail and file sharing. The computer organization basically remained an "octopus," however, even if some of the tentacles functioned on their own.

■ THE NETWORK FLATTENS CORPORATIONS

If the personal computer sparked the revolution in corporate culture, the local area network (LAN) won it. Instead of being subservient to a huge mainframe, each computer in a network could assume an equal role in the system. And if each computer was equal to the others, so was each computer user. Aside from the eventual impact on the way applications evolved, the most important role of the network was to flatten organizations, creating a new culture in the process.

Corporations everywhere have been cutting out their middle levels of staff, but that wouldn't be possible without the network. It's created a way for executives and managers to inspire, direct, and supervise more than the traditional dozen or so reports. Every morning, every worker around the world can receive the same message from the CEO about progress in meeting sales goals or new benefits. Every assembly worker can learn about important changes in process. There's little need for a pyramid of management—a pyramid that all too often distorts messages as in a children's game of "telephone."

It's not a one-way channel, either. An individual worker can tell the CEO about a problem that is costing the company money and time, and the CEO can get it fixed in short order instead of having a suggestion crawl up through hierarchies and committees, perhaps for weeks or months.

The flat organization inevitably encouraged individual contribution and responsibility, but it also fostered teamwork. Teams, however, may be composed of individuals in many departments in many locations, not just people whose desks happen to be close together. The cottages are virtual in many cases, not real.

As networks changed corporate cultures, they also changed organizations and processes. Personal computers and sophisticated technical workstations became more powerful, and they assumed more of the burden of the enterprise, soon far eclipsing the mainframe in computational power and impact. Not that the mainframe disappeared—it was usually kept to maintain its legacy applications such as accounting and payroll or to become a more sophisticated information server.

Today it's difficult to distinguish between a personal computer, a server, and a mainframe simply on the basis of their raw computing power. Servers with capability little greater than desktop PCs run huge applications that once required mainframes, if they were practical at all.

This type of computing is called *client/server computing*. The client—the individual desktop computer—does much of the work, accessing the server only to get information it needs or to store it for future use or for other users. It allows each person or organization within a corporation to contribute its part fully, whether that contribution is the product of a single writer or a whole accounting system.

➤ Harnessing Client/Server's Power

The move to client/server computing liberated users and departments in corporations, but like most liberation movements it created a certain amount of anarchy. The old mainframe systems were easy for their operators to control. They could easily restrict access, prevent changes, and prohibit certain operations. With hundreds of computers sprouting around a company, however, MIS and in effect the company, lost control.

The response by MIS was to try to harness the power of all of these individual computers in a more organized manner. It turned out to be very difficult. Dozens, sometimes thousands, of different applications were responsible for functions throughout corporations. Most served one function or user without any thought of sharing work or information with others who might need it.

The eventual solution was to create suites of software modules that can serve virtually all of a corporation's needs, operate in concert, and contain the security and controls vital to running a company, yet users have their own powerful applications. Today's corporate information systems are based on two complementary technologies: client/server computing and relational database managers. Companies such as SAP, Computer Associates, People-Soft, and Baan are working to perfect and install these packages, some for specific industries, some usable for any corporation or organization. These suites are programs that "sit" on powerful relational database managers from companies such as Oracle or Informix, accessing and using the data in many ways.

Modern database managers can hold more than simple data, in fact. They can store complex *objects* consisting of both data and programming instructions. These objects could be video, the complete description of a 777 wing, or a program to calculate life insurance risks. Using a relational database is not for the timid, however—hence the need for specialized, prewritten applications such as SAP financial modules.

That doesn't mean that there's no need for programming, but it's normally done by specialists with high-level tools provided by vendors. Consulting companies help install these enterprise applications, and this consulting has become a big business for the management consulting arms of the big accounting firms and many other experts.

These experts, in fact, are virtually required for a company to install or convert to enterprise systems. The task is too big for in-house MIS people to learn on the job, and the systems generally require significant customization to meet a specific company's needs. Once installed, however, the systems are relatively easy to use and maintain. They typically include programs that make it easy for business specialists to exploit their capability without programming, simply by interacting with the data in a familiar Windows graphics interface.

■ MAKING PROCESSES FIT THE SOFTWARE— OR VICE VERSA

One of the biggest challenges for most corporations that decide to install an enterprise system is matching corporate processes to the software. That's not necessarily because the software demands that the company work in a unique way, but because most companies operate in a manner based on history and tradition, not always efficiency. The first step is to evaluate the company's processes and work flow, which will almost certainly change as improved efficiencies of integrated systems eliminate and simplify many steps. Thus the company doesn't just buy and install software; it's forced to evaluate and perhaps reengineer its whole business. The software only facilitates this process, making change possible.

This customization, however, shouldn't mask the strong tendency for corporations to buy expertise, rather than develop it in-house. As is true everywhere else, corporations are buying services and products that help them run their businesses rather than developing their own tools. In the past, most companies hired flocks of programmers to create programs in COBOL to meet their needs. Now the companies buy off-the-shelf packages and customize them with relatively user-friendly, higher-level tools. They've learned, as the late John Welty, general manager of Motorola Semiconductor, once said, "We've learned that if we can buy tools, it's cheaper than building our own—no matter how much it costs!"

■ THE INTERNET OPENS THE FINAL FRONTIER

One of the most significant developments in technology's transformation of corporations is the impact of the Internet. It's having an enormous impact both inside corporations and beyond. The logical extension of the local network, the Internet is the tool that lets the individual connect to the world.

Popular attention is focused on the Internet's impact on individual consumers, where it has vast potential for entertainment, education, communication, and shopping. All of these functions attract business interest, and both new and existing firms are likely to benefit from serving these needs.

Bill Gates is one of those who changed his whole company's strategy when he saw how important the Internet would be. "Year by year, the Internet is driving into the mainstream. The user interface will be dramatically better. The bandwidth will be much better. More and more people will be used to it. Kids who go to college in the United States today live what I call a Web lifestyle: for any major thing they want to do—plan a trip, make a purchase, coordinate things with their friends—they're going to use the Internet as part of the process.

"Even if they don't actually place a car order through the Internet, they'll know before they go into a car dealership exactly what the dealer paid for that car, so they'll certainly be better at negotiating. And so as more and more people move into the workforce, as we get the ease of use and manageability to be far better, slowly but surely electronic mail and all these information systems will become the mainstream."

Gates continues, "I think it's safe to say that within 10 years the majority of all adults will be using electronic mail and living a form of that Web lifestyle. And they'll begin to ask, 'Why can't I file my taxes that way?' Well, of course, you will be able to. 'Why can't I ask my doctor a question or schedule an appointment easily electronically?' Some professionals will jump at that opportunity and make that possible."[5]

The Internet could transform businesses in other ways, too. Using its basic technology in the form of an intranet, the Internet creates a superb method to interconnect and inform employees while it's drawing them together.

Organized as an extranet, the technology lets vendors, buyers, resellers, and customers join in one virtual enterprise, allowing them to access and share information, even engage in electronic commerce. Both intranets and extranets are simply subsets of the Internet, sometimes only a click and a password away from the whole Internet and its resources. Those, too, are transforming many corporate functions, as using the Internet replaces old ways of doing things, sometimes creating new resources.

Lewis Platt, chairman, president and CEO of Hewlett-Packard, says, "In the last few years, we've seen several waves of change in the computer industry—moving from mainframe computers to minicomputers to PCs. HP has played a major role in each of these waves, from the minicomputer days of the HP 3000 to being the first major player to embrace open systems to our worldwide leadership in client-server and UNIX systems.

"The Internet is the latest wave. Like many of the other waves, it's associated with chaos. It's not an evolutionary step, it's a revolutionary step. It's a disruptive technology. I believe that the rise of the Internet is the biggest development in our industry since the invention of the telephone.

"We're seeing the creation and development of a global infrastructure for information. It's never existed before. It's developed very rapidly. This creates tremendous opportunity: Because of the Internet, we can now implement solutions that we've only dreamed about in the past."

Platt adds, "This chaos presents a tremendous opportunity. We believe our role at HP is to help bring the two worlds of PCs and systems together synergistically to create a new category of products and solutions. But as these systems become more pervasive, we have to redefine the phrase 'mission critical.' Historically, we've used the phrase to describe a few applications that were important to the company. They were somewhat compartmentalized, and lived separately in a 'glass house.' In the new world, with the development of the Internet, the entire network and infrastructure become mission-critical."[6]

■ CONTROVERSY OVER THE ROLE OF THE NETWORK COMPUTER

Ironically, the Internet technology that liberates individuals could also partially restrict them once again. The technology creates the possibility for a partial return to centralized computing, albeit a much more palatable form than in the days of dumb terminals.

When a personal computer user launches a Web browser, the user is turning this PC into an extension of the computer at the other end. The local computer can execute programs, but it's often asking another computer somewhere to do the work. That means that any browser on any machine is compatible, and it also means that the local computer doesn't really have much to do except display graphics and text and issue commands. This is ideal for finding information and conducting research, but not necessarily for analyzing and manipulating data.

A number of companies have proposed simple network-oriented PCs that can be used on Internet-type networks. Such PCs wouldn't necessarily have fixed or removable local disks and would be completely dependent on server computers for storage and manipulation.

For many applications, this concept makes great sense. The network computers ("thin" clients, as opposed to "fat" PCs) would also be cheaper and easier to maintain, since they could be upgraded at the server. For call centers, support desks, and many other applications where the users basically access and record information, these computers would be ideal. Many MIS directors also like the idea that they could recover control, and they could determine the applications for which the computers could be used. That might include keeping users from surfing the Web or playing games at work.

Network computers wouldn't be ideal for anyone who needs to manipulate data locally, however, or where communications links might be a bottleneck. Notes Microsoft's Gates, "I think companies need to think about the environment they create for their knowledge workers. Knowledge workers are not like factory equipment. They're not cogs in a process. When you give them empowering tools, when you let them see the information, when you let them have the context of knowing what your company is planning to do and how their job relates to that, it can make an incredible difference."[7]

The network computers don't represent a full step back, but are more likely to be used to upgrade present terminal applications while giving users far superior graphics, usability, and access. They're likely to find wide use in industry, but not replace conventional personal computers and workstations that need local storage and power.

One vendor who is skeptical of the role of the network computers is Mitchell Kertzman, chairman of database vendor Sybase. "There's a mutual conspiracy between vendors and customers. Industry tries to sell a magic bullet to fix problems, often one they created," jokes Kertzman. "Client/server computing was supposed to cure the ills of host-based computing, but just when we're getting to the hard part, managing all those fat clients, vendors are claiming that thin clients will fix the problems. So instead of making the products we've sold work, we sell a magic bullet."

Whatever the fate of network computers, "Weblications" are here to stay. Serious business uses of the Internet have become important, and they'll become more so in the future. It's difficult to imagine any corporate functions they won't touch, as the following chapters will show.

■ TECHNOLOGY IN THE CORPORATION

Today's corporation couldn't exist without technology, and this book examines how companies use technology to improve productivity and efficiency. The heart of the book features examples derived directly from business professionals using technology, from end users to vendors and to observers of technology. The examples are not intended to be comprehensive, simply illustrative. There's also no attempt to draw sweeping conclusions or predict far into the future. The evidence speaks for itself. It's inconceivable that the information revolution will fail, for it's already occurred and there's no turning back. Our corporations and our lives will never again be what they were in the past. Technology has changed them all forever, and it would take a bona fide Luddite to call the changes bad, either for individuals or society.

Former nontech CEO Gerstner says it well: "I came to IBM as a customer, believing that information technology would transform every institution in the world; that it is one of those technologies that comes along every century that truly revolutionizes society and every part of society. And being inside IBM, I am more and more convinced that this is one of those change-the-world technologies."[8]

A special note: Almost every example in the book mentions the impact of the Internet, so it's appropriate that this book is partly a child of that technology. Much of the research, most of the examples, and many of the contacts were initiated over the Internet. It's the most exciting development in communications since the telephone, and it's sure to have as big an impact.

Chapter

1

Turning Concepts into Products

■ INTRODUCTION

Time has taught us that even the largest enterprise can't excel at every task. As companies reengineer to focus on their strengths and to source out other services, many realize that product conception, development, and realization are their prime sources of differentiation. And the ability to be first to market with a new concept may be one of the greatest factors in success and profitability.

Technology is used in the product development process in many ways, from facilitating research to creating models to the complex engineering needed for most modern products.

The ideal way to conceive of clever new products might be to find a modern-day Leonardo. This is obviously difficult, but there are many ways in which technology can help lesser humans develop similar reputations. For the actual process of designing the product and turning it into something that can be manufactured reliably in volume at minimum cost, however, there is no substitute for modern technology. Early two-dimensional computer-aided design (CAD)—basically, automated drafting—led to static three-dimensional representations, then to models that maintain their characteristics even when they are rotated on screen, almost like holograms.

Now designers can even create moving three-dimensional products. Such virtual reality images can be seen through special goggles that allow the customer, for example, to view cars from different angles or even "sit" in them and see if the radio tuning controls are easy to use. And, the ultimate designer's dream, these electronic models can be automatically turned into prototypes using remarkable replication machines to build plastic models that exactly duplicate the objects displayed on the screen.

Once proven, these designs can be sent to a factory, where computer-controlled tools and test equipment reduce the need for human intervention as well as the possibility of errors. Some products literally could not be developed without computer assistance. The design of the modern microprocessor, for example, extends far beyond the ability of any group of people to implement—unlike the early integrated circuits, which were laid out using X-Acto knives and sheets of opaque plastic that acted to mask certain photosynthesized areas from light.

■ FACILITATING PRODUCT CONCEPTION

There are many ways that companies conceive of new products. A few, like Xerox Corporation, have set up independent groups to create new types of products. Xerox's justifiably famous Palo Alto Research Corporation (PARC) came up with many product ideas—the personal computer (PC) as we know it, the graphic user interface (GUI), laser printers, the mouse, and the Ethernet local area network (LAN)—though it never successfully exploited most of these developments.

Few products are truly unique, of course. Except for the rare flash of brilliance, such as the Post-it Note from 3M, most new products evolve from existing products—either those of a particular company or those of its competitors. Today's Internet technology makes it fairly easy to find out, as a starter, what's available.

An idea, however, is a long way from a finished product. Even before undertaking serious development, many companies now conduct extensive research to determine whether a need exists that would justify the effort. Perhaps the most common form of research for this purpose is the *focus group*. Basically, this is a group of people, selected as typical potential customers and positioned behind a one-way window, observing a facilitator who tries to evoke impressions about product concepts or prototypes from the group. These days, the sessions are almost always videotaped, and sometimes they are televised over special networks to remote locations. More commonly, key items are edited to present to the managers who control the purse strings on development.

Most large corporations have used focus groups in developing new products, notably respected suppliers of consumer products such as Levi Strauss and Ford. Many companies conduct other types of consumer research, including suppliers of ergonomic office furniture such as Herman Miller and Steelcase, which have on-site labs and conduct studies at customer sites.

■ VISUALIZING PRODUCTS

Once a product is approved, development begins. The first step, whether in automated or manual design, is *visualization*—what the product will look like. Second comes the actual product design, followed by prototyping and possible changes, creation of manufacturing tools, then production and testing. Few companies have yet integrated the whole process, much less completely tied it into other corporate systems, but that is obviously the direction in which many want to head.

Engineers have been using computers to develop products for years, but some of the most exciting applications are in visualization, a way in which firms such as Chrysler and Boeing are significantly reducing development time and expense. The first step in visualization, whether for a cereal box or an aircraft, is generally a relatively simple concept sketch. More and more, designers create them with a computer rather than on paper—it's much easier to modify and reuse electronic images. Artists frequently sketch with a paint program, which deposits tiny colored dots on a screen to create a picture. These are generally in two dimensions—basically, an electronic version of paint and paper.

Drawing programs differ fundamentally from painting programs in that the designer creates lines and geometric shapes rather than painting dots. These programs can produce more compact files, and they can also be enlarged and reduced without the distortion that comes from the same process with a painting program.

Sophisticated versions of these drawing programs were the first computer-aided design software. Really computer-aided *drafting*, these programs allowed the drafter to draw lines and other shapes very precisely, to correct mistakes and modify content, to copy and reuse shapes, and to store and print drawings. Like word processing, this process has become so efficient that traditional drafting tools are becoming as obsolete as slide rules or typewriters.

By its nature, this type of CAD is two-dimensional, but in real life almost all products must be constructed in three dimensions. At first, these 2-D images were used just like paper blueprints. The artist prepared various views, which helped others visualize the item and were used to create manufacturing tools such as molds and jigs. High-quality rendering can produce what looks like a photograph, and it can be printed out or transmitted electronically to anyone involved in the process.

These drawings by themselves can't be directly used in product development and manufacturing, however. They can only suggest the form of a product. To generate more useful drawings, various programs allow users to convert these sketches into more sophisticated drawings or even 3-D objects. This may be done manually, depending on the subject. These three-dimensional objects can be rotated for viewing from any direction, and they can actually store all the dimensional characteristics and finishings of a real product. They are so good that, in many cases, using a computer means that a company doesn't have to sequentially make dozens of prototypes to find the best design. This can result in a huge savings in money; more significantly for many companies, it substantially reduces the time it takes to get a product to market, a critical factor for success or failure in today's fast-changing world. Products as simple as perfume bottles and as complex as automobiles receive these previews.

For more and more companies, in fact, the digital description is the product specification. These companies are abandoning paper drawings as the standard for digital models. For example, Boeing has reached that point in its massive design for the 777 aircraft (see Chapter 2).

➤ Designing in 3-D

Even simple products can be improved by using CAD. Pfaltzgraff Company in York, Pennsylvania, is one of the nation's leading suppliers of fine ceramic dinnerware, maintaining the tradition of handcrafted quality that began 200 years ago. The firm offers more than 60 patterns of formal and casual dishes that combine beauty with a handcrafted look. In the late 1980s, the company wanted to streamline its design process, in particular to produce supplemental items such as serving dishes that could be added to its existing lines but would sell in lower quantities. These pieces had to have the same look and feel as the original pieces, but Pfaltzgraff also wanted to be able to produce them more quickly. Some patterns required that the random-looking, subtle "throwing rings" of the originals, an intentional design element, be reproduced. CATIA, the leading solids-modeling software, was chosen because it was able to introduce these rings, making the pottery look handmade.

To create these as well as new pieces, a designer now works with a CAD technician who inputs the instructions on an IBM RS/6000 RISC workstation. The CAD operator not only enters the original design, but adjusts the surface texture and tool design, as well as programming for the numerically controlled machines that produce molds. Unlike some types of products, Pfaltzgraff creates only one master mold, which is used to produce the working molds for manufacturing the pottery. For that reason, speed in making molds is less important than accuracy.[1]

Samsonite is famous for innovation, and its products are designed using the same CATIA software that Pfaltzgraff uses. Although it seems to be a low-tech product, in truth the luggage is quite sophisticated. Samsonite's Epsilon line, for example, though designed to be distinctive looking but inexpensive and durable, includes cushioned handles, combination locks, shock-absorber wheels, piano-type hinges, and tongue-and-groove mating surfaces to keep contents from spilling. The case is made from polypropylene, the same tough plastic used in automobile bumpers.

Samsonite says it developed the products in only three years rather than the five years typically required in the past using hand-made wooden models. And this reduced time frame was for the first products the company developed entirely using the 3-D software. Even the prototype injection molds were created directly from the software, allowing the company to bypass expensive sculpted models. Like Pfaltzgraff, Samsonite's CAD operators enter data from designers. At some companies, however, the designers enter their ideas directly.[2]

A refrigerator door doesn't seem very sophisticated, but Frigidaire uses 3-D software to create its products, too. As consumers demanded new features, the firm found it increasingly difficult to depend on the old 2-D drawings of the past. The different views and renderings were simply not accurate enough. Since the company went to 3-D solids modeling, however, there's no longer room for misinterpretation, which often led to expensive engineering changes when parts didn't fit or the products didn't look the way the designer intended. As with Samsonite, Frigidaire's toolmakers make their molds from accurate electronic models, not mock-ups. The company has been able to eliminate the iterative process of building prototypes, testing and refining them, then repeating the process until everything was right.[3]

Black & Decker goes even further than Frigidaire, allowing product designers at its centers around the world to design their own parts on the computer, then share designs, even working concurrently on projects that can be successfully merged. Like many other firms, B&D started using the CATIA 3-D software as a simple 2-D drawing tool, as the company did with the CADAM software it had used previously; however, B&D then moved to using the 3-D model as the definitive representation of the product rather than the old 2-D drawings. For Black & Decker, this is especially important as it tries to expand the capabilities of its small electric tools. Now the dimensions and placement of the products are better defined.[4]

➤ The Move to Solids Modeling

The pivotal step in the wider application of CAD was the step from 2-D drafting (basically a more efficient version of the old drafting T square, triangle, and pencil) to solids modeling. Simple in concept, its implementation awaited powerful workstations before it could be widely applied.

Early 3-D programs required mainframe computers or supercomputers plus sophisticated graphics terminals. Today's engineering workstations, even many of today's high-end personal computers, can far exceed the power of those old products. That means that many workers, not just a few specialists, can use them for design and development.

Many of the earlier so-called 3-D programs could really only store an outline, or "wire frame," version of a part. It wasn't practical to add surface textures or change lighting as the part was rotated— itself a slow process even for just the simple wire frame.

Now 3-D solids modeling programs such as CATIA, Parametric Technology's Pro/Engineer, and Alias/Wavefront design and visualization software store all the details of an object, allowing it to be rotated yet remain realistic. Because the interior can also be modeled, companies like Black & Decker can peel away the surface to work with the engine inside.

One company that learned from experience how much computer power it takes to use these programs is Outboard Marine Corporation, manufacturer of Evinrude and Johnson motors. Outboard Marine uses large design drawings, requiring accelerated graphics to display properly. The company also generates thousands of pages of documentation, and both text and graphics must be shared, often with remote locations. It's a demanding application for computers.

As Outboard Marine was developing its new 25- and 35-horsepower Mirage motors, it found that its computer-aided design system was slowing to a halt. The problem wasn't the software, for the company had invested heavily in Parametric Technology Pro/Engineer and Mechanica design aids that performed their tasks well. However, it found that its hardware wasn't up to the job. The company's engineering department was especially concerned about an upcoming task: merging five separate, large data files to create a design for the engine block. It decided to try Silicon Graphics Indigo2 workstations, which were optimized for computational- and graphics-intensive applications. That solved the problem. The new system had an added advantage: Superior graphics allowed the company to eliminate physical prototyping of parts (depending on a model) which were now generated using stereolithography instead. This process took three days and eliminated many months out of the design cycle. All in all, Outboard Marine says it cut cycle time by 30 percent, speeding time to market, while producing a higher-quality product.[5]

One appeal of the electronic design process is that many companies find they can create computerized modules that can be combined to form new products. Why design a part from scratch when something similar has already been designed and perhaps even exists in physical inventory or production? This is as true for a maker of glassware as it is for making aircraft at Boeing—though, admittedly, the number of modules needed for a jet aircraft is larger.

Such parts can be more accurate than the physical ones, in fact. Computers are obviously good at math—that was their original application—making this an easy task. For complex products, computer-aided design allows many people to work simultaneously on the same project. This includes documentation, manufacturing, and testing—processes that once had to be done in sequence.

➤ Automobile Firms Depend on 3-D CAD

Apart from electronics companies themselves, the biggest early converts to CAD/CAM (computer-aided manufacturing) were automobile and aircraft manufacturers, who made products so complex that they could hardly be built without the aid of computers. CAD/CAM systems received perhaps their biggest endorsement when Boeing designed and built its 777 jet without any physical prototypes, an enormous change from the past when hordes of employees made many physical mock-ups, a process that sometimes continued for years.

Even making a clay model of a car requires many employees and about three months' time. Now a single designer can do the same work in a few weeks and provide variations (such as different colors) to boot. Ford estimates that using 3-D technology will cut nine months from the approximately three years' development time for a new car beginning from concept, saving the giant carmaker about $200 million per year.

An indication of the importance of this software is that Ford spent $4 million to standardize Alias/Wavefront software, scrapping its internally developed computer-aided industrial design (CAID) processes. Alias/Wavefront software is an important part of Ford's C3P project, intended to integrate its design, engineering, manufacturing, and product information management systems into a single electronic interface that will be used worldwide by Ford and its vendors.

Ford has already launched vehicles using C3P and the company expects all of its future cars to be produced using the system. It has also recommended that its suppliers adopt the same technology standards to improve communication throughout product development and manufacturing.

The Alias/Wavefront software provides Ford with an integrated path from initial 2-D concept sketches to 3-D models. It also provides integration with the company's CAD/CAM systems. It takes a number of different products to accomplish the whole job, however. Ford is already the world's largest user of Alias StudioPaint 3D, and now the design teams can easily integrate their 2-D sketches with 3-D models built with Alias AutoStudio. Using Alias I-DEAS Direct Connect technology, Ford engineers can export geometric models created in AutoStudio into its Structural Dynamics Research Corporation (SDRC) CAD/CAM/CAE (computer-aided engineering) software. The whole network lets Ford integrate its entire design process while facilitating collaboration between concept car stylists and engineers all over the world who convert the ideas into automobiles.[6]

Like Ford, Chrysler is a big user of CAD technology. Long before its cars are built, they're visualized and designed on Silicon Graphics workstations, which create images so real they look like they can be driven. The process allows Chrysler to develop and deliver better-designed cars more quickly than ever was possible before the company committed heavily to 3-D technology.

Chrysler had been using an internally developed CAD program called CAD/CAM, plus various commercial products, when it acquired American Motors Corporation in 1984. Chrysler found that AMC's Jeep subsidiary was using an early version of the CATIA CAD software. In 1989, the automaker reorganized its engineering division along vehicle platform lines. At that time, the company standardized using CATIA as its single, core CAD/CAM system to maximize efficiency and integration. It also adopted the CATIA Data Management (CDM) system to make engineering and release information available to all corporate users. This step helped the company release its LH series of midsized cars six months ahead of the original schedule. The Jeep Grand Cherokee was the first Chrysler vehicle to have large portions—fully 50 percent—designed in CATIA. The Ram truck introduced in 1993 was 90 percent designed in CATIA, and other cars are now developed completely using this system.

Obviously, many elements make up the automobile design process at Chrysler, from the glamorous overall look, to engine development, and even to the perfecting of small components such as rearview mirrors and door handles. The thread that ties the whole system together is CATIA Data Management, which handles the prosaic but daunting task of keeping track of all the parts and models so that they can be retrieved to make "digital cars."

Before CDM was implemented, Chrysler had stored the models for components under the designer's computer name. To get the model, the user had to know the designer's name—no problem when the system started, but a huge nuisance as the company's roster grew to include thousands of designers. This was an especially troublesome problem for users of the engineering data (such as manufacturing and documentation departments). The company says that some people spent as much as 60 percent of their time simply trying to find the right data. It was a great waste of time and resources, and increased the time Chrysler took to get new cars into the market. Now designers simply retrieve a part number when they begin designing a part. Others who need the information can get it from the CDM database, which is cross-referenced in many ways. Many different engineers can simultaneously access the information and download it for their needs—whether creating tooling or equipping a new car with a previously designed and tested component.[7]

The CDM system also contains information about how pieces fit together—hardly a trivial matter in a product as complex as an automobile. This is accomplished by dividing a new car into zones, with one engineer responsible for each zone. This information is used to retrieve data for many purposes, including Chrysler's sophisticated digital visualization and prototyping programs.

As the component models enter the database, engineers and designers download them to Silicon Graphics servers and workstations to create digital cars. This process has allowed Chrysler to develop cars more quickly, even while allowing more design iterations. The company used to make prototypes out of Styrofoam and wood, a process that could take 12 weeks and required highly skilled workers. If the parts didn't fit, they had to be modified or even made over. Today, the company just checks them out using its internally developed Chrysler Data Visualizer software on the workstation. "Now, we create digital prototypes in minutes," says Chrysler's supervisor of advanced systems development, Hugh Cumming. "We can run interference checks that tell us immediately if we have a problem. If we do, we can resolve it in minutes as well. As a result, we have eliminated many mock-ups and prototype parts." This has eliminated the need for many repetitive mock-ups and prototype steps.

Not all the work focuses on the look of the cars. Using sophisticated software that also requires high-performance computers, Chrysler engineers perform demanding rigid dynamic analyses on the prototypes, checking wheel and axle movement and suspension and handling, as well as other characteristics. The company can also run simulated crash tests to see how a car will react in an accident. Nevertheless, the biggest impact of having powerful computers and software is that very realistic simulations can be created very quickly, allowing more design iterations and more complete engineering and tests.

The last step in visualization is the most dramatic, because, at this point, designers, engineers, and managers use virtual reality systems to really appreciate the car's look and feel. Donning an immersive display helmet and hand motion tracking gloves linked to the computers, the "driver" can evaluate the car's interior. "We use virtual reality to check for ergonomic factors—can the driver easily read the dashboard displays or easily reach the radio and temperature controls—as well as for issues such as good visibility," says Cumming.[8]

Once the car passes the test, information is deposited in the CDM database, where it is accessible to manufacturing, documentation, and other departments. The CATIA 3-D model, in fact, is the master part description for all product development and subsequent activities at Chrysler, including those of the shop floor.

Use of the software combined with a rethinking of the whole product design and engineering process has helped reduce to 32 months the time it takes to develop a new car. "CATIA was a major factor in saving as much as one year of development time on the Neon," says Annie Guillou, strategic account manager at Chrysler. It has also helped Chrysler become one of the most cost-effective automobile producers in the world. Chrysler's experience with CAD/CAM, in fact, goes beyond designing vehicles. Its success with cars has led it to use similar techniques to redesign its facilities for greater productivity.

➤ Designing a Big Cat with CAM

Many seemingly old-line businesses have strongly embraced advanced technology. Caterpillar, the world's largest manufacturer of construction and mining equipment, natural gas engines, and industrial gas turbines and a leading global supplier of diesel engines, is one. Its classic yellow equipment can be seen at most highway construction sites. But at its Mapleton, Illinois, division, Caterpillar is designing and manufacturing the parts you don't see.

This division works with design engineers to build pattern and core boxes for the manufacture of a wide range of parts for anything that needs an engine, from stationary power units to tractors to earthmoving equipment, including blocks, heads, liners, and CAMs. The group uses Unigraphics 3-D software for all its modeling and manufacturing. The parts that this Cat plant manufactures are so complex that they demand a flexible modeler, because an extreme amount of filleting and blending is required. There are no vertical walls or straight lines in the engines, so 90 percent of the design is free-form.

Says pattern designer Jess Stanfill, "An engine requires many threaded and drilled holes that run from top and bottom. Each hole needs a sphere or boss (a male sphere) built around it to provide for stock. With the bosses going in all different directions, there is no design latitude—the geometry is very restrictive. The pattern makers can't change a curve. Therefore, Cat uses free-form surfaces over solids."

The large size of Cat's assemblies also adds to the complexity. The engine block of some of Cat's engines is so large that children could play in it. Yet a cylinder head has literally thousands of features and there are hundreds of parts in the bill of materials.

The Mapleton division does both new design and redesign. It's easy to import legacy data into Unigraphics software, and the product designer can also communicate with other Cat plants, some of which use a different system of specifying drawings.

Regardless of the application, there is usually a difference between the perfect answer and the real-world solution. In the real world, engineers do not just enter numbers in an equation to come up with a perfect answer or perfect math model, they apply real-world knowledge to find the best solution that gets them where they want to go. For Caterpillar's designers, that real-world knowledge means using a flexible, free-form model from Unigraphics.

Cat uses some plastic and wood prototypes for certain applications, but more recently the company began going completely digital. When doing work for internal customers such as other Cat plants, the Maplewood division can do it digitally without using physical models. Cat's concurrent engineering direction supports going directly to the final product from the digital file.[9]

➤ A Virtual Locomotive

Almost symbols of the past, locomotives are probably the last product most people would expect to be created with modern, advanced virtual product development programs. However, locomotives represent a $1 billion market, and it's a lot easier to create a 415,000-pound, 80-foot-long locomotive on a computer screen than it is to build a prototype of it! General Motors' Electro-Motive Division (EMD) in LaGrange, Illinois, does just that. Using EDS Unigraphics' Virtual Product Development (VPD) capabilities, EMD has made significant inroads toward realizing mass customization and rapid time to market. This has led to dramatic dividends for the division, including a 50 percent savings in design and drafting time as the result of configuring its basic locomotive to accept three different engines. This has also led to significant savings in the cost of building a new locomotive.

"I would like to have a 3-D model of a locomotive on a workstation that I can take to the customer, modify during a live session to suit his requirements, then transfer directly to the engineering office and from there to the shop floor," says Dick Dunteman, general director of engineering at EMD. It makes an enormous difference to the project that the customer can see what the unit will look like and that it will meet all the requirements of the contract beforehand.

One of EMD's major projects is what it calls its "platform" or basic locomotive. EMD has taken advantage of Unigraphics' modeling capability to provide the base locomotive starting point. Most of the critical parts have been entered into EMD's 3-D database, and the full digital mock-up in Unigraphics lets EMD analyze and make changes quickly, with confidence that things will work right the first time out. EMD is also working on a revolutionary 6,300-horsepower engine. No one in this industry has ever delivered a brand-new engine from a "clean sheet of paper" in three years, but the division plans to achieve this through VPD.

Working closely with Unigraphics, EMD sees its move toward full virtual product development depending on four core competencies: digital product and process modeling, product information management, computing and communications, and organizational and process change. By linking all analysis, design optimization, and machine-tool and fixture programming to modeling, EMD expects a significant reduction in prototyping time. By doing the up-front analysis simultaneously with creating the model, the division believes it will be able to final-test the reliability of the engine in much less time than it has taken in the past, even with more stringent reliability goals. For EMD, the keyword has been *digital,* to enable live design sessions and high-speed transmissions. EMD wants to stay away from paper and physical prototypes to achieve the goal. The division needs a digital model of the product and process, it needs to maintain that model on-line, and it needs the computing, communications, and organizational infrastructure that will enable that.

EMD's digital master model process relies on technology such as 3-D solids modeling, top-down assembly modeling, attached business attributes (e.g., material type or supplier name), and smart assemblies based on advanced representation. Likewise, *associativity*—tying different files together—plays an important role, with drawings based on master model views; an automated bill of materials; integration with CAE, CAM, and quality measurement; technical document integration; and supplier integration. With its digital database, EMD is now addressing product data management through complete vaulting of data and is expanding its system to address work-flow and configuration management.

The locomotive development process revolves around a colocated team of engineering and manufacturing people working simultaneously on product design, analysis, manufacturing, fixture design, and serviceability around the virtual model. An engineering group in London, Ontario, also works from the same 3-D database. The two groups conduct simultaneous design sessions live at both locations, transferring information back and forth over a high-speed telecommunications line. The division also transfers data electronically to remotely located manufacturing shops for the machining and pattern work.

Obviously, standards are vital when an organization is geographically widespread. EMD has a wide area network (WAN) that includes its UNIX workstations. A consistent office environment and a high-speed communications link between sites contribute to smooth electronic communications, including video sessions sharing live CAD. Similarly, the company's supplier integration program has benefited from the network.

Successfully rolling out new technologies and reengineering processes has made EMD, originator of the diesel electric locomotive, a new technology leader in the transportation market.[10]

➤ Physical Models from Virtual Designs

For all the advances in computer displays and virtual reality, however, there's a need for physical models of many products. For some, such as food boxes, it's easy to print a prototype that can simply be folded and pasted to make a model of the eventual mass-produced version. For most products, however—even simple bottles—there was no practical way to make a solid model automatically until recently. Some could be tooled with a numerically controlled lathe or milling machine, but this process can be prohibitively expensive. Now corporations such as Chrysler and Logitech as well as services including DTM Corporation and 3-Dimensional Services have sophisticated machines that can produce 3-D models—some relatively inexpensively.

These machines produce physical models using a variety of technologies. Some are very precise. A number of techniques build thin layers to create parts. Three-dimensional printing uses a technique much like ink-jet printing to squirt a binder in successive layers of metal, plastic, or ceramic powder, which are deposited on a piston that slowly retracts. When the part is finished, the excess powder is brushed away; it also serves to support overhanging or hollow parts during the printing. The process called *solid ground curing* builds up thin layers from a light-sensitive polymer that solidifies, leaving the unexposed area liquid so it can be removed. *Laminated object manufacturing* (LOM) uses a laser to cut thin layers of paper that are laminated to form the part. *Selective laser sintering* (SLS) uses a laser beam to bond nylon, metal, or other powder into the shape of the part.

All of these techniques seem to be straightforward industrial processes, but another one seems to come from a science fiction script. The *stereolithography apparatus* (SLA) process uses a laser beam to solidify a part in a vat of liquid polymer, like an alien oozing from the earth. But it works well.

Some of the parts produced by these technologies are used primarily as conceptual models, while others are suitable for making molds for producing plastic or metal products. Still slow and expensive, they can cost $100,000 to buy and take days to produce complex, large models. Nevertheless, it's often much less expensive and faster than alternatives such as hand-sculpting or machining.

Eastman Kodak, for example, has used SLS from DTM Corporation to model components for cameras, copiers, and other products. Some parts are created using automated tools; some are sculpted by hand in clay and then digitized with the 3-D version of a scanner to create a 3-D file. This process was used to create a polycarbonate prototype of a camera face using SLS, for example. Kodak says that the SLS process made a significant impact on the company's product design and development.[11]

Porsche used the same SLS process to create a prototype metal cylinder head in just 4 weeks instead of 16, and at a cost under $12,000 instead of the $75,000 required for the sand casting it had previously employed.[12] Hewlett-Packard used SLS to produce two pairs of complex parts for a plotter prototype at about $6,000 in five days rather than the $26,000 and 1,200 hours required to make them with numerically controlled machining.[13]

ABB Nera, a Norwegian supplier of mobile satellite communications systems, used a solid ground curing machine from the Israeli company Cubital to produce a prototype case for a new briefcase-size phone in only hours after the CAD files were completed. Fourteen plastic parts were made, plus three molds that were subsequently cast in aluminum.[14]

The world of satellite communications changes rapidly, but it can't compare with the fashionable market for sporting goods and clothes. Smith Sports Optics of Ketchum, Idaho, used stereolithography to develop high-tech ski goggles that would fit better and allow better peripheral vision than existing models. The company shaved three months off its total design cycle, including studying three iterations in a two-week period, something impossible with traditional model-making processes. The products were so accurate that the company could use them to produce finished-looking samples for trade shows and retailers.[15]

The ultimate goal of these model-producing technologies is to be able to produce working molds or patterns—even final products—directly, perhaps using compressed metal powders. However, with working times of hours needed to produce simple shapes with expensive raw materials on more expensive capital equipment, industry is still a long way from the time when it can turn out custom solid shapes as if they were personalized T-shirts.

■ DESIGNING THE MICROPROCESSOR

Developing an expensive, complex, and large product such as a car or airplane using computer-aided design techniques may be impressive, but for sheer wonder, perhaps no manufacturing process is more remarkable than those that produce tiny integrated circuits such as microprocessors. Like parents teaching their child, in fact, these products are used to produce their successors, often stretching the capability of even the latest microprocessors to develop the next generation.

The invention of the transistor by researchers at Bell Labs ushered in the modern world of electronics, but it was Bob Noyce's development of a method for making practical integrated circuits that created the information explosion of today. Noyce developed a way to isolate and interconnect individual transistor switches and amplifiers on a single, tiny slice of semiconducting silicon material. This step led to the fingernail-size circuits of today, which now contain literally millions of individual transistors composed of elements as small as one-sixth of a micron (a millionth of a meter) on a side.

The process used to make integrated circuits includes a number of exotic procedures, all performed with almost unimaginable precision. The basic idea is to "dope," or infect, certain areas of nonconductive silicon with ions that make it either conduct or block electricity, depending on the electrical charge placed on the area by small conductive plates. The dopants or impurities are introduced either by turning them into gases that migrate into the silicon at high temperature or by shooting them into the crystal structure. The little switches, in turn, are interconnected by tiny ribbons of aluminum, in many cases stacked like highway overpasses and separated by thin, nonconductive silicon oxide (glass) areas. These shapes are defined primarily using multiple sequential masks that alternately block and allow deposition, define metal routes, and permit or inhibit etching of silicon, glass, and metal.

The process used is *photolithography,* also used routinely in offset printing of stationery and business cards, but the minute dimensions and multiple steps make it probably the most demanding manufacturing process the world has ever seen. Originally, chip designers defined and laid out the required patterns on large sheets of drafting paper, then skilled technicians created large masks for each of the required layers using opaque sheets of plastic that were cut by hand with a sophisticated version of conventional manual drafting tools. These were then reduced tremendously and projected in precise patterns on the surface of the chips to expose layers of chemicals that either hardened or could be washed off like offset patterns.

Early chips had only a few layers—perhaps 8 to 12, but even these required superhuman attention to produce parts that worked the first time, after months of development and as much as a few weeks of processing. As parts got more complex, steps in the process were automated—mask patterns, layout, entry of logic patterns, predesign of whole logic functions, generation of test patterns, and so forth."

➤ Making Custom Chips Quickly

Automation makes it possible to produce modern integrated circuits. Without automation, there is no way that today's complex chips could be built. Even so, the process can be very time consuming. It is not unusual to take six months or more to produce a chip to a customer's design. That's simply unacceptable in an industry that introduces whole new computer or communications products in that amount of time.

One of the secrets to quick turnaround in manufacturing chips is microcontrollers—tiny, special-purpose computers that manage everything from garage door controls to machine controls. The average car today contains 50 to 75 of these microcontrollers, controlling everything from the air bags to the engine itself. Being miniature computers, microcontrollers can be programmed in their software to change their properties, making the same basic product useful for a wide range of applications. This software, however, has to be embedded in special areas of the chip, so custom design and manufacturing are required. For high-volume applications, it is desirable to integrate the microcontroller brain with other functions on the same chip, which is an even more complicated matter. Because chip size is so critical in volume applications, these functions are handcrafted to optimize die size, a function at which humans still excel.

Many firms, including Intel, Hitachi, and Zilog, make these microcontrollers, but the industry leader in microcontroller chips is Motorola, with 30 percent of the market. Like other chip makers, Motorola has been attempting to reduce response time for customers so that they can get their products on the market quickly. Says Murray Goldman, executive vice president and assistant general manager of Motorola's Semiconductor Products Sector, "Our customers are driven by cost, low power and features, but the issue of how to do things faster is probably the most fundamental point of competition today."

Chips that are designed to meet customer requirements are called *customer-specific integrated circuits* (CSICs), and, until the early 1990s, it took about seven months to produce them. Realizing that this was far too much time, in 1992 the company undertook the major task of replacing its existing design process and technology, choosing an integrated suite of design tools from Cadence Design Systems, the leader in the chip design market. This effort paid off by reducing the cycle to seven weeks by 1994—a significant improvement, but one that whetted customers' appetites for a further reduction.

It was obvious that a significant decrease in cycle time would require more than an improvement in the existing design process. It would take a completely new approach. The old process started with existing designs, which were pared down and redone as needed to fit new functions. The process was laborious and prone to errors, although the end result was an improvement over the past.

Motorola's new process, called "7-day CSIC," starts with a clean slate, and it is designed to produce new chips in only seven days. Rather than depending on the memory and insight of individual engineers, the new approach uses a fully documented, consistent process. It also uses Cadence software tools, but in this case the two companies had to work as partners to completely reengineer the way Motorola approached producing these chips. The parts are better designed and are often smaller, but the big improvement has been to remove many of the steps that had previously led to delays.

As an example, the company used to devote to the task two or three full-time engineers, plus two or three layout technicians, for six to eight weeks; now two or three people are required for just a few days. A specific example is the task of designing the highest-level schematic diagram and interconnecting all the required modules, plus getting all the connections right and eliminating errors, which used to take someone one month. Now it takes ten minutes.

The project has had other benefits as well: In addition to reduced labor, reduced use of equipment has resulted. In the past, the company might have had to run a simulation 40 times to eliminate all the bugs from a chip. Now it has to be run only once, saving a huge amount of computer time. Employees are liberated from tedious drudgery and rework, too, allowing them to address more creative tasks.

All in all, the project met its goals—and even bettered them, with some chips being designed in as little as four days. Having already seen significant improvement, Motorola expects to pay off its expensive investment in a year. More to the point, the company would have become noncompetitive if it hadn't made the investment. Customers like to see the improvement in products and cycle time, and they also value Motorola's commitment to their businesses.[16]

■ DEVELOPING NEW FOOD PRODUCTS WITH COMPUTERS

Not all product development is as mind-boggling as that of microprocessors, but some is as exacting and technical in its own way. One example is the nutritional content of food. Consumers—and the government—are taking an increasing interest in the ingredients in foods, including the nutritional values of packaged food. At Best Foods, a division of CPC International, this translated into a lot of work. In the past, the company's researchers laboriously calculated nutrient totals for each product, using food composition references in the U.S. Department of Agriculture handbooks. This was a slow process, particularly when using the firm's old, time-shared computer system, and also because the company's research staff regularly updated formulas for current products, line extensions, and new developments, which necessitated new content analyses. The staff also had to calculate content for the consumer recipes the company published as a promotional effort. As is probably apparent, food product companies are introducing more and more items, and competition is growing, so there is pressure on the firm to speed up the process of content analysis.

Best Foods found a solution: With a New Jersey systems integrator, Spectrum Concepts, the company developed a nutrient databank system using Informix software. The database is a computerized list of over 6,300 food items and ingredients, each having up to 74 specific nutritional values, plus 2,000 recipe and product formulas. The database is used to calculate the nutrient profile of recipes and product formulas, retrieve nutrient information for food items or ingredients, and simplify the preparation of ingredient statements for products. The software automatically calculates content, cutting in half the time it took to prepare the statement.

Best Food developers and researchers can also quickly determine whether products align with nutritional preferences, such as those for low sodium, low fat, or low cholesterol content. And, the staff doesn't need any programming expertise—anyone in R&D can easily use the system.

Informix software is used in many other areas of product development at Best Foods. For example, it's used to track recipes developed by test kitchens, to manage resources, to report R&D project status, to conduct periodic analytical product quality testing, and even to track job openings for the department.[17]

■ TECHNOLOGY IN PRODUCT DEVELOPMENT

Technology has come to play a large part in product development and design, from assisting in research to eliminating drudgery. Its biggest role, however, is computer-aided design, which is used for everything from creating initial concepts to producing realistic 3-D models of products that increasingly serve as the definitive descriptions of parts and products. Computers are also vital tools for engineering products, too, including sophisticated programs that help lay out and simulate integrated circuits. Aside from providing descriptions for manufacturing, however, product design departments remain isolated enclaves at many companies and have largely resisted the integration efforts that are impacting so many other functions.

$\mathscr{C}hapter\ 2$

Making Operations Operate

■ INTRODUCTION

If this book had been written 20 years or even a decade ago, this chapter on manufacturing and operations would have been very different than it is. In the past, corporations focused their efforts on improving productivity, primarily in the area of automation. Removing human involvement from manufacturing was the big objective, both to contain costs and to improve quality and repeatability. Among the hot issues were *computer-aided manufacturing* (CAM), *computer-integrated manufacturing* (CIM), "lights out" factories containing no people, and robotics.

With technological advances, these issues became less important. Suppliers and users have developed and perfected equipment that can automate virtually any task. Whether to employ such equipment in any particular case is more of an economic decision than one of availability, except in some areas, such as making semiconductors or even soldering tiny chips on circuit boards, where the manual dexterity of humans is not precise enough to be practical.

There's hardly an area of manufacturing, especially in electronics and automobile production, where automation hasn't made a strong impact. At a time when even toasters contain microcontrollers and the average car boasts 50 to 70 of them, this means that automation affects almost everything used by business, government, and consumers.

Now the battleground has shifted to other issues: company organization and work flow, information systems, outsourcing, building parts to customer order. Having wrung operational costs—if not capital spending—out of manufacturing, companies are taking a step back to figure out what to do next to become more competitive.

As in the fast-changing functions elsewhere in the corporation, even the current concept of manufacturing may be in the process of becoming obsolete. Some of the biggest "manufacturers" in American industry farm out much of their products—either buying and reselling someone else's product with a new label or color, or designing products that someone else manufactures to order.

At one time, the sign of a "real" electronics company was the huge flow-solder machine that it used to produce circuit boards. Now even fabled electronics suppliers like Hewlett-Packard and IBM have sold major production facilities to other companies, only to contract with those same firms to manufacture the computer components and assemblies they once built themselves. These companies even sell some finished products made completely by other firms, although they always have a strong hand in the design and quality of such products.

At the same time, the whole idea of what constitutes a vendor has changed. Vendors have become valued partners, not just entities to squeeze for the lowest possible price.

Companies no longer fill warehouses with inventory in the hope that it will sell, but they wait until they receive orders—or even until they deliver products to customers—before building the product. These companies have learned that technology and market demand change so rapidly that doing things the way they used to could leave them stuck with a mountain of white elephants.

Fortunately, technology—and a new way of looking at business processes—is making it feasible to build to order. Current technology makes it practical to assemble a single product exactly the way a customer wants it—even a very complex product like a truck or computer system—and ship it to a customer in days instead of weeks or months.

In many ways, it's meaningless to separate a company into functions when these functions are no longer really distinct. Purchasing was often *part* of manufacturing in the past, but now it may *be* the core of manufacturing. And with the increasing acceptance of integrated, enterprise-wide information systems such as those from SAP and PeopleSoft, information technology decisions and operations could become more critical than any other single issue in a corporation.

■ RECORDING NEW PROFITS

New approaches to manufacturing are having a big impact on many more companies than just the more obvious ones, such as those dealing in computers and cars. Even record companies are changing the way they make products. EMI Records Group is one of the largest music companies worldwide with annual sales over $2 billion and 15 percent of the market. EMI is also a manufacturer; it actually makes the CDs it sells.

In the 1990s, dramatic changes have hit the music business. First, the whole distribution system for music has consolidated, with big chains like Blockbuster and Tower grabbing more and more of what had been the province of small retailers. At the same time, development costs have been skyrocketing, as expensive music videos, concert tours, sponsorships, and other promotions replace the old, low-key efforts of sending free records to radio station DJs. Performers are demanding more money, too.

Fortunately, new and more accurate information sources are allowing music companies to better understand what customers want, and if these companies can respond, they benefit greatly. The new sources of market information include data from point-of-sale systems installed by those big record retailers and monitoring of radio airplay. Both methods are immediate and accurate—far more valuable than yesterday's statistics gathered after the fact from telephone surveys.

A giant like EMI, with its many hot artists, should have been able to take advantage of these trends, but it suffered from the system it had been using. First of all, EMI's logistics for distributing CDs to retailers were inefficient—and expensive. EMI ended up with a lot of obsolete products, and, worst of all perhaps, had a bad reputation among its customers for service.

It was apparent that this wasn't an isolated problem, but that the whole operation needed to be reworked. Senior Vice President

Joe Kiener initiated a study, hiring Computer Sciences Corporation's Index group and its Consulting & Systems Integration unit to lead a team of EMI experts from across the company to look at the problem. It discovered a number of problems.

To control costs, EMI had traditionally maintained the capacity to manufacture about 80 percent of its own CDs, relying on outside firms to handle the overflow in periods of high demand. The company was paying millions of dollars for this work, yet it constantly ran into problems with shortages and excess inventory. In the past five years, however, the cost to replicate CDs has dropped significantly. There wasn't much penalty for having excess production capacity, so EMI increased this capacity from 80 to 120 percent. This allowed the company to completely eliminate outsourcing and it increased on-time delivery from 90 to 99 percent.

The company still had a big problem with stock, however. It had three distribution points, and one was constantly running low, while the others had excess inventory. "The old paradigm was that all SKUs (stock-keeping units) had to be treated in the same manner, and that all three warehouses had to carry everything," says Kiener. Research among retailers determined that they had different expectations of delivery for different types of CDs. They needed hot albums immediately, but with older, "catalog" items, they were more tolerant. In addition, demand is different for "rare" CDs than for those old ones with continued wide popularity.

Eventually, EMI reorganized its warehouses so that one holds the entire line of 10,000 units and the other two warehouses distribute the 500 SKUs that accounted for the products in most immediate demand—and 85 percent of sales. After these changes, EMI increased customer service while cutting costs significantly. Now manufacturing is in sync with distribution—and ultimately with customer demand.[1]

■ REORGANIZING TO BETTER MAKE BIG AIRPLANES

Since the 1940s, when the Boeing Company pioneered mass production of large airplanes, its tradition of engineering and technical excellence have made it one of the world's most successful manufacturers of large jet airplanes. Boeing's engineers used the latest technologies to develop innovative airplane components and systems driven by the needs of its customers. The company's competitive dominance, however, was based more on the technologies that were applied to airplane component design than on efficient processes and systems.

By the late 1980s, Boeing realized that the systems it used to manufacture airplanes had not progressed as quickly as the technology itself. After several isolated attempts to improve processes, the company realized that it needed to revolutionize the whole way it did business. After studying other companies around the world, Boeing concluded that it had to be flexible enough to evolve as the market changes, while reducing production costs and defects by 25 percent and time to market by 50 percent.

In 1994, Ron Woodard, president of Boeing Commercial Airplane Group, initiated a strategy that attacked several key problems. The resulting project, called Define and Control Airplane Configuration/Manufacturing Resource Management (DCAC/MRM), simplifies the processes and systems used to configure and define the components for each airplane, to build and assemble those parts, and to manage the data.

DCAC/MRM is a cooperative effort between Boeing and its information technology providers. It integrates suites of off-the-shelf software into a highly flexible and powerful system that allows the firm to select the right process for each job. The program helps the diverse organizations of the large company function like an integrated network of small businesses.

Before DCAC/MRM, Boeing's size reduced its efficiency. It had grown and split into separate organizations with distinct products, goals, and objectives. The business processes and interlocking information systems of these organizations evolved to meet their own needs, not those of Boeing overall. For example, the company used more than 800 computer systems to design, manufacture, control, and support airplane components and assemblies. Most of these systems had narrow functions, were not integrated, and did not communicate with each other. To make matters worse, some were at least 40 years old.

Complicated processes and systems required new design and planning every time the company built an individual airplane, even if 75 percent of its components were common to the model. The bill of material created for each airplane had to be converted or manually reentered as many as 14 times.

A special headache was a configuration control system that assigned a customer-specific identification number to each engineering part drawing. Each drawing was marked manually with this number even if the same part had been built and used a hundred times. The drawings were even re-marked when airplanes changed position in the production line, to ensure accurate customer configurations. Not surprisingly, this redundant labeling slowed production, wasting 95 percent of the time spent re-marking documents.

Two elements of the new program changed that. The Define and Control Airplane Configuration (DCAC) element allows Boeing to configure airplanes more efficiently, and Manufacturing Resource Management (MRM) improves the way it builds the planes. DCAC/MRM has four key elements: a single source of product data, simplified configuration management, tailored business streams, and tailored materials management. Each element substantially reduces costs, cycle time, and defects.

The DCAC/MRM system architecture and software supports up to 45,000 employees using 40,000 workstations at 77 sites. All the information needed to configure, define, produce, and support Boeing airplanes is represented by a single source of product data, replacing 410 internally developed systems once used throughout the company.

The DCAC/MRM system is based on open standards, client/server computing, and commercial off-the-shelf software applications. To create the new system, Boeing worked with Baan, SRDC, Trilogy, CIMLINC, Iona, Oracle, Hewlett-Packard, and Sequent.

The firm replaced its cumbersome configuration control with an automated system, called the *airplane-specific configuration table,* which tracks parts according to airplane identification. This allows Boeing to provide components just in time for installation, using a single bill of material for each airplane rather than the multiple bills of material required in the past.

The firm uses the single source of product data and the airplane-specific configuration table to separate components of each airplane into three categories, or tailored business streams. The first includes basic components common to all airplanes of a major model. The second comprises those components common to all airplanes of a minor model that have been previously delivered and certified. The third comprises unique assemblies that Boeing hasn't previously designed, manufactured, or certified.

Boeing's tailored materials management lets it choose the best materials management process for each production situation. Previously, the firm used the same complex and costly process for every part. Separating work into business streams allows Boeing to choose the most effective processing and scheduling method. For example, Boeing can schedule a basic and stable part using less complex and less expensive methods than those needed for a newly defined one that has not been built before. Separating components in this way allows the company to take advantage of commonality among airplanes of the same model.

DCAC/MRM also allows employees to access data much faster than before. People share information and recognize how their work contributes to the company's business. When a customer or a supplier asks about a component that involves more than one manufacturing or assembly unit, a Boeing employee can answer quickly by accessing the single source of product data. Sales representatives use the system to help customers understand the factors that affect cycle time, weight, and cost of producing an airplane.

DCAC/MRM represents a radical overhaul of the way Boeing uses information technology. With this system, the company is able to integrate complex, flexible data sources so users can find information quickly. This flexibility allows its business practices to evolve in a constantly changing environment.

➤ Combining Off-the-Shelf Software to Create a Custom Solution

The systems integrated by DCAC/MRM can change rapidly in response to external issues, such as market requirements and government requirements. The challenge was to integrate software applications that had not been designed to work together into a cohesive, easy-to-use system using a loosely coupled integration approach. Every software package maintains its individual integrity. Boeing melded the packages by creating standardized "plugs" that allow it to remove or upgrade one package without disrupting the whole system.

Distributed-object technology was the key enabler. It evolved during the early 1990s when an industry and end-user consortium, the Object Management Group (OMG), formed to create specifications and to motivate key players in the computing industry to adopt this technology. It is built around *objects,* reusable software blocks that can be developed indpendently, stored and cataloged, then combined to create new applications.

Applying the OMG Common Object Request Broker Architecture (CORBA) led Boeing to an innovative system integration approach that helped it design a reliable information system with great flexibility. It was one of the first end-user members of OMG and the first to investigate and prototype the use of distributed-object technology in a large-scale, complex manufacturing environment.

The firm used early prototypes to define an approach in which custom systems could be assembled rapidly from standard components, most of which are off-the-shelf software. The company identified a set of core services common to many distributed systems and not directly related to any particular business process. These services and a commercially available object request broker became the application integration layer. It also developed well-defined, reusable interfaces based on CORBA specifications, which provided a standard to work with.

DCAC/MRM incorporates proven software packages designed by suppliers to respond to the requirements of many different companies. The evolution of information system technology allows Boeing to leverage that technology and apply it to its manufacturing environment on a massive scale. Because it can now make use of this advanced technology, the company is taking a significant stride toward a goal of reducing by 50 percent the time between initial negotiations with customers and the delivery of a finished airplane.

Boeing's partners in DCAC/MRM share the manufacturer's commitment to reducing complexity and simplifying the manufacturing process. Baan provides the software that manages the overall flow of parts, the inventory levels, and the financial transactions. SDRC provides a product data management tool that can access the single source of product data. Trilogy provides software that makes it possible for sales representatives to work anywhere with airlines to configure and define their airplanes. CIMLINC provides software used for work instructions and to plan the process by which the part is built or assembled on the airplane. Iona provides the enabling technology that allows Boeing to integrate the other software applications. Oracle supplies the database, and Sequent provides the clustered database servers. The applications reside on more than 300 Hewlett-Packard servers at 77 work sites throughout the company. DCAC/MRM is the largest project of its type in the world. It will eventually be deployed on 40,000 workstations and 300 servers in support of more than 45,000 users.

The company relies on an internal network or intranet as a critical tool for ongoing support of the project. It uses the network to collaborate on development efforts, share electronic files, provide documentation, and deliver training. It also conducts more than 30 percent of its business process training through this type of communication.

DCAC/MRM has become the foundation for Boeing's business and production strategies. Where it has been implemented, the company has seen immediate and impressive results. The initial sites, at its Skin and Spar manufacturing facilities in Puyallup and Auburn, Washington, went on-line in February 1996. Data following implementation at these sites, which employ 700 people, indicate that DCAC/MRM significantly reduced the defect rate and increased the speed of production. By the end of 1996, Boeing had implemented DCAC/MRM in six factories and in three support organizations, while re-deploying every person displaced by its efficiencies.

The company shifted to computer-aided process planning that provided higher-quality work instructions, augmented workload mobility, and modified plans and manufacturing strategies. For example, a single airplane-specific configuration table eliminates hundreds of thousands of drawing sheets. Before implementation of DCAC/MRM, up to 14 bills of material were needed for each airplane to support the work flow. All these bills of material were entered separately, initially were only from 60 to 70 percent accurate, and required much rework. After DCAC/MRM was implemented, all processes used a single bill of material that was 99.7 percent accurate at the very outset, according to Boeing.

Boeing also found that it could quickly determine the total cost of any airplane component at any point in the manufacturing process. For example, although the wing of a 747 is one of the most complex airplane components, the system generated a total cost breakdown of the piece in less than 15 minutes. This was the first time the company could achieve such accurate cost visibility.

In another example of time and cost savings, a customer decided to change a cargo door on a 777 to a larger cargo door. Documenting the change of a cargo-door size would have taken 5,087 hours under the old system. Using DCAC/MRM, it took four hours. Because of dramatic time savings, Boeing was able to incorporate the change much earlier in the production run. Only the first airplane down the production line had to be retrofitted, saving a significant amount of money for the customer.

The greatest challenge Boeing faced in implementing the system was to completely change the way it produces airplanes. Many of the processes and systems had been in place for 40 years. Its staff was used to them; they were traditional, but they were no longer efficient. The implementation of DCAC/MRM directly affected 45,000 people in Boeing Commercial Airplane Group and impacted virtually every other organization in the company. Consequently, it created a great deal of anxiety. The company had to make sure that it communicated these changes clearly to everyone involved, from production employee to senior executive. The nature of the product dictated that the company develop extensive training programs that would withstand the scrutiny of regulatory agencies. The company's information technology partners worked closely with Boeing to train employees and implement the new system.

Another challenge was to prepare 3,100 airplane component suppliers for the new processes and systems. This took a focused communications program, and the firm continues to offer specialized symposiums for suppliers from around the world to ensure their understanding of the new processes.

Because DCAC/MRM will evolve with the changing business environment, Boeing is faced with the ongoing challenge of developing new ways to communicate, train, and implement a constant flow of improvements. It is no longer dealing with a static, traditional system, but with one that grows with the world around it. The company's dynamic relationship with its manufacturing suppliers and its contining efforts to interact effectively with employees and customers will ensure that Boeing enters the new millennium using the best, most efficient, and most rewarding business practices.[2]

■ USING ENTERPRISE SOFTWARE TO TRANSFORM MANUFACTURING

Boeing had a Herculean task set before it: re-engineering a system using 40-year-old processes (and in some cases technologies) required to manufacture probably the most complex product ever built. A more usual scenario is that of 3Com Corporation, a far newer company that makes products that have entered the marketplace more recently.

3Com, a leader in data networking, offers a broad range of products and systems such as routers, hubs, and adapters that form the foundations of global data networks for its eight million customers. 3Com is replacing its legacy information systems with enterprise-wide client/server applications that integrate its order-entry, order-fulfillment, accounting, materials management, manufacturing, and distribution operations around the world. The new system utilizes SAP's R/3 application software, Informix-OnLine Dynamic Server, and Informix-ESQL/C, running on Hewlett-Packard computers. This technology strategy improves 3Com's customer responsiveness, lowers its cost of sales, and increases its ability to respond to dynamic market opportunities.

3Com has thrived by responding to its customers' needs for products that reduce the cost and complexity of data networking, creating products that span the needs of functions ranging from desktop computing to wide area networks. In 1991, the company launched a strategic initiative to expand sales and earnings by capitalizing on its global capabilities, building market leadership through new products and aggressive marketing, forging strategic partnerships, and streamlining operations. This required restructuring the company to focus on the emerging global data networking market and rebuilding the company's product portfolio to sell complete integrated network solutions, rather than point products as it had in the past.

3Com began actively acquiring technologies and companies to supplement its product lines. It extended its global reach by opening a European manufacturing facility and design center, a new service center in England, and new sales offices around the world.

Keeping pace with these strategic business changes and growth proved too much for the company's mainframe-based logistics system. The company's information staff was constantly modifying its system just to keep up with changing business needs. Eventually, the system became too labor intensive and costly to maintain for a $1 billion business and when 3Com decided to internationalize its business, it became clear that the existing system was simply unable to accommodate cross-functional and intercountry integration.

After extensive evaluation, 3Com decided to replace its legacy information system with one based on open systems and relational database technology. This would provide 3Com's worldwide operations with on-line access to critical business information and enable it to capitalize on standards-based applications and productivity tools to lower its information costs. 3Com says it chose the SAP R/3 application suite partly because of its outstanding support for international business. It also selected Informix as the system's underlying relational database foundation.

With the assistance of Price Waterhouse, the system integrator on the project, the new Informix-based system was implemented in two phases. The first concentrated on integrating global finance, sales, distribution, and some areas of materials management and production planning for 3Com's U.S.- and U.K.-based worldwide logistics centers. The new system provides real-time access to manufacturing and logistical information, enabling order-entry clerks to check credit and stock availability, as well as authorize product shipments and pricing at order time. In most cases where the product is in stock, it is shipped the same day, eliminating unnecessary time and costs from the order-fulfillment cycle. This greatly facilitates inventory management for 3Com and its customers—and gives 3Com a market advantage over its competitors. The system also lets 3Com use new technologies such as electronic data interchange (EDI) to further shorten the order-fulfillment cycle.

Phase two extended the system's integration to include manufacturing, accounts payable, and fixed assets, giving 3Com a fully integrated, enterprise-wide decision-support system. The new system enables 3Com to fine-tune its business to meet diverse international demands. The system simplifies international sales by accommodating various sales structures, pricing schemes, and product allocation strategies, while ensuring that 3Com's global needs are maintained. The system will also facilitate 3Com's worldwide distribution and planning by enabling 3Com to plan and execute production in the company's supply centers. With half of 3Com's sales and a quarter of its staff based outside North America, this is of great value for the company.

3Com's order-fulfillment department utilizes PCs to access a network of application servers and Informix database servers across a high-speed fiber-optic network. 3Com's 200 concurrent users process about 1,000 order-related transactions per day, with orders varying in size from as few as 10 to as many as 20,000 products. The system is expected to grow by one gigabyte of transaction data per month. OnLine Dynamic Server's scalability and performance will enable 3Com to easily expand the system to accommodate its growing business.

In addition, 3Com expects to convert its sales and marketing data warehouse to OnLine Dynamic Server. The data warehouse maintains sales order invoice shipment information, as well as product part serial numbers and sales details on the millions of parts 3Com sells each year. These serial numbers are captured on the manufacturing floor when the part is created and will be tracked by the application against actual orders and shipments to facilitate customer service, logistics, and planning. In the future, as 3Com's workforce becomes more mobile, 3Com expects to exploit Informix's support for mobile users and leverage the system's remote access capability by providing customers with dial-in access to order information—in their own language and currency. 3Com also expects the Informix-based system to enable it to continually reduce cycle times in manufacturing and order fulfillment, helping to increase its productivity per employee and reduce manufacturing and logistics expenses.[3]

Another company that turned to SAP was Hewlett-Packard, the industry's second-largest computer supplier. HP, too, decided to undergo the complex process of updating its systems as its business grew dramatically and became increasingly like a consumer business in its volatility, volume, and price structure. Like many other firms, HP had a collection of diverse software programs developed in-house. And because of its worldwide operations, they were incompatible and varyied widely in capability. They certainly didn't meet the company's growing needs, especially for greater flexibility, shorter lead and delivery times, and supply chain–oriented order entry.

Also, like other companies in the same situation, HP found that it had to change its whole business process but decided to make the changes and introduce the software gradually, starting in its German operation. This group first rolled out the materials manufacturing, production planning, and parts of the financial accounting and controller functions.

One complication was that HP had to develop about 50 links to its custom software. The tool it developed to do this eventually became a standard HP product. Eventually, it took about 18 months to implement the software for 8,000 products. After it was well proven, HP rolled SAP out to other divisions around the world.[4] Likewise, number-one competitor IBM is installing the same software suite, starting with production applications at its Storage Division, a major supplier of disk drives, with five manufacturing sites in four countries. The division uses it to conduct business in more than 40 countries around the world. Eventually, the software is expected to be used throughout IBM.[5]

➤ Making a Cosmetic Company Look Better

Though SAP has gotten the most attention among the software companies that are helping customers reengineer their operations, many companies have chosen other routes. Nu Skin is one. Nu Skin is a $500 million U.S.-based personal care and nutritional products company that uses a multilevel distribution strategy to market its products.

Nu Skin was founded on a simple premise: An innovative company dedicated to producing the highest-quality products could attract a loyal clientele within the highly competitive traditional personal care and nutrition market. Nu Skin's premise has been proven correct. Today, the company has 100,000 distributors in the United States alone and has opened new markets in Canada, Taiwan, Hong Kong, New Zealand, Australia, and Japan.

But selling product to the marketplace is only one challenge. Delivering product is another. A few years ago, Nu Skin relied on a proprietary mainframe computer for sales ordering and a nonintegrated DOS PC-based program for inventory control. Sales forecasting was done in an ad hoc manner. The result was that the company had little control over its inventory, and product shortages occurred on an almost daily basis.

Nu Skin's marketing approach has created tremendous growth and this placed heavy demands on its manufacturing and distribution operations. To better forecast sales and control inventory so it could avoid product shortages, Nu Skin installed an enterprise-wide solution consisting of an integrated suite of materials requirements planning (MRP), distribution, and accounting applications based on Informix database technology.

The applications from Inter-Data Systems (IDS), a value-added reseller for Informix, have enabled Nu Skin to accurately forecast sales opportunities and expedite the movement of materials and products through its warehouse. The result is tighter control over inventory and a much leaner and more efficient distribution system. The company saves at least $2.2 million per year in shipping costs alone.

"We simply lacked control over our manufacturing and distribution processes," says Jeff Hunt, manager of corporate business systems at Nu Skin. "As a multilevel marketing company, Nu Skin's primary focus is its distribution system: We rely on our independent distributors to both sell products and to find new distributors. When you have distributors selling at a fast pace, but then you can't supply the product because you didn't forecast sales accurately, you lose business and your image is tarnished."

With the goal of dramatically improving its sales forecasting and manufacturing process, Nu Skin considered a number of options for a new information system. The company quickly determined that it wanted a relational database for flexibility and one that was based on the UNIX operating system for cost efficiencies. After evaluating solutions from over 40 different vendors, it selected the IDS Promax suite of integrated applications that run on the Informix-OnLine database and were developed using Informix-4GL.

"We were attracted to the Informix/IDS system because of its superior price/performance over any of the other solutions we looked at. It provides full integration of our accounting, manufacturing, and distribution processes," says Hunt. "It gives us complete control over each phase of our operations, and expedites processes by eliminating redundant activities. With more than 125 on-line users, we can transfer data easily among modules, which streamlines operations and provides greater accuracy."

Today, Nu Skin's sales orders are still entered into the company's mainframe computer. From there, however, the information is downloaded each morning to the Promax master production control module. This module forecasts sales using both firm sales orders from the mainframe and projections based on historical and seasonal activity. The module then sets up schedules needed to produce a certain number of products by a certain date. Once the schedules are arranged, this information flows to Promax MRP II, its material requirements planning module. This module checks existing inventory as well as raw materials in stock, then calculates raw materials needed to complete orders in the forecast.

With Informix/IDS, the company has a fully integrated system. The system controls Nu Skin's sales forecasting and is proactive in determining the inventory needed in stock. It has also improved its quality control and has a much better handle on its manufacturing process.

This level of control allows Nu Skin to move product through its manufacturing process more quickly than before. Nu Skin has efficient warehouses that handle between 4,000 and 8,000 orders per day. Everything is bar coded—products as well as pallets. Using handheld scanners, workers scan the bar code. The information is then transferred via radio frequency to a receiver, which then sends it over the network to the Informix database on the central computer. That makes inventory information accurate because it's scanned and stored in the database. Scanning also allows Nu Skin to move products quickly because it takes only seconds to scan a pallet of goods.

The Promax manufacturing and inventory applications are also linked to Nu Skin's Promax-based accounting modules—accounts receivable, accounts payable, general ledger, and payroll. As product is shipped or supplies are ordered, data flow into these modules automatically. This provides Nu Skin with current and immediate information, which allows the company to bill more quickly and pay its vendors promptly, improving vendor relations and letting it take advantage of vendor discounts. With multimillion-dollar orders, 2 percent discounts add up to a savings of between $250,000 and $500,000 per year.

Nu Skin entered a market dominated by large, established companies and successfully carved a lucrative niche through innovative distribution and commitment to quality. Now it has the technology to meet the escalating demands for its products.[6]

➤ Applying Exotic Technology at Deere

While EMI, Boeing, 3Com, Hewlett-Packard, and Nu Skin chose process redesign and proven technology to significantly improve their manufacturing, another major company turned to exotic new technology to accomplish that objective. Deere & Company leads the agricultural equipment industry as an innovator in applying information technology to manufacturing. It also actively pursues technology transfer from commercial, academic, and government organizations to supplement the skills of its information technology staff. This approach shortens the time it takes to implement new ideas, helps control development costs, and introduces its staff to emerging technologies.

The company licensed OptiFlex, a genetic algorithm–based optimization system, to automate assembly-line sequencing at the John Deere Harvester Works, a factory that makes row crop planters. This first commercial installation of OptiFlex was also the first practical installation of genetic algorithm technology for scheduling commercial manufacturing. Genetic algorithms are probabilistic search techniques that take an initial set of feasible solutions and improve them over successive generations using mathematical methods based on principles of population biology such as selection, crossover, and mutation.

In 1992, Deere Harvester Works introduced a new line of row crop planters that radically changed planter production methods. Previously, Deere had filled orders by selecting components and options from factory inventory for shipment to local dealers who assembled the products. Now, it assembles the planters at the factory and ships them directly to the dealer so that they never become finished-goods inventory. The improvement in factory production, finished-product inventory cost, and delivery time has justified the cost of changing production methods.

Unfortunately, this is a complex process, since each product sold can differ significantly from others. OptiFlex addresses the extremely complex problem of producing all models of planters on a single assembly line. Since the daily model mix changes continually, creation of an efficient assembly sequence requires the scheduler to balance competing goals of manufacturing and marketing and resolve complex production and resource constraints simultaneously. Time constraints limited Harvester Works schedulers to producing weekly rather than daily schedules. Substitute and replacement schedulers often required extensive on-the-job training to become proficient.

Deere began a search for information technology to enable Harvester Works to generate daily schedules, release the scheduler for more strategic planning and coordination tasks, and cross-train other individuals to produce schedules.

Deere identified the new technology—constrained optimization using genetic algorithms, which has been successfully applied in military and space applications—as potentially useful for automating its assembly-line sequencing. Constrained optimization integrates the competing objectives of manufacturing and marketing into a total factory solution. Genetic algorithms employ a random, yet directed, method that produces better schedules than traditional analytic methods. Deere chose Optimax Systems Corporation, a new company with expertise in these technologies, to develop a prototype scheduling system for evaluation. Optimax demonstrated a successful prototype within three months and Deere installed a production system, which integrated an improved version of the prototype scheduler with other plant systems.

The OptiFlex scheduling system enabled the Harvester Works to improve its capability to schedule orders for assembly daily rather than weekly, revise customer orders closer to the build day, meet proposed build schedules more consistently, ship more efficiently, reduce finished-product inventory and the associated inventory carrying costs, and balance competing manufacturing and marketing goals.

Customers do not want to choose from a number of features suggested by the manufacturer. They only want the features they want. Constrained optimization using genetic algorithms can enable the various manufacturing practices required by this dynamically changing market. Soon, all successful manufacturing system vendors will provide constraint-based planning and scheduling functionality, and optimization engines will follow next. By taking advantage of the previously existing information technology, Deere & Company and Optimax Systems Corporation have demonstrated the practical benefits and future potential of this approach for others to follow.

OptiFlex technology has its origin in work done under government R&D contracts. Optimax Systems Corporation was formed in 1993 to pursue commercial application and to continue development of this technology for manufacturing applications. Optimax knowledge engineers created the prototype OptiFlex constraint model from the scheduler's rules of thumb and interviews with production supervisors, then tailored their proprietary genetic algorithm search for assembly-line scheduling applications.

Deere & Company funded the first commercial application of the OptiFlex technology to a discrete manufacturing scheduling problem. Deere believes this risk was justified, since the cost of development was repaid in savings during the first two years of operation. Five other Deere & Company factories, plus appliance, heavy truck, and automobile manufacturers, as well as a major health care provider across the United States and Europe, are also adopting it.

The opportune linkage of a business need and appropriate information technology made this innovative application possible. Harvester Works introduced a new line of row crop planters and radically changed planter production methods, while Optimax Systems Corporation founders were successfully applying OptiFlex technology to military and space applications. Deere & Company staff, charged with improving order fulfillment, introduced the parties and funded development of the prototype. In this case, the benefits of the application of genetic algorithms resulted from participation in Internet newsgroups, corporate membership in the Santa Fe Institute Business Network, and collaborative research with the Center for Complex Systems Research at the Beckman Institute of the University of Illinois at Urbana-Champaign.[7]

➤ Even Small Companies Benefit

It's not just giants like Boeing and Deere that benefit from improved manufacturing control, however. Tiny Skutt Ceramic Products in Portland, Oregon, makes kilns for hobbyists, schools, potters, and specialty tile manufacturers. It builds its ovens from the ground up, including grinding bricks, cutting sheet metal, installing heating elements and manual and electronic controls, assembling, testing, and shipping.

For 14 years, the business had been using an IBM System 23 similar to a PC. After 8 years, it added a Novell Network, later adding MAS ADD+ON accounting, distribution, and manufacturing software on a Hewlett-Packard NetServer. Skutt's personnel appreciate the information available on-line. When an order is entered, the system automatically creates a work order, thus saving time. A bill of materials for a kiln comprises about 200 items, including all the nuts, bolts, and screws. While this is a repetitive manufacturing environment, Skutt adds job-shop components, such as different voltage and distributor stickers, to meet customer needs.[8]

➤ Thomson Uses Advanced Manufacturing Technology to Build Dish Systems

Not all the improvement in productivity comes from process and software, however. Some still comes from better manufacturing technology. It has helped Thomson Consumer Electronics become a leader in consumer satellite systems. Thomson, part of the global Thomson Group, is the world's fourth-largest consumer electronics company and the largest television manufacturer in the United States.

The way Thomson did it, however, illustrates how different manufacturing has become from the past. Thomson didn't want its staff to design, build, and install a whole manufacturing plant. It wanted to hire an expert to do the job.

Headquartered in Paris, Thomson designs, manufactures, and markets color TV receivers and picture tubes, VCRs, camcorders, audio and communications products, and the RCA brand Digital Satellite System (DSS). DSS consists of an 18-inch satellite dish, a digital receiver, and a remote control. The integrated receiver/ decoder for DSS is assembled in one of Thomson's four manufacturing facilities in Mexico.

Thomson looked for a total system solution. It wanted a company to take its requirements and implement the project from layout to complete installation. Thomson's challenge was to find a system integrator that was willing to accept the responsibility of integrating equipment from multiple vendors. It found that most vendors wouldn't take on the challenge—they wanted to deal only with their own products.

Besides value, Thomson needed a vendor with a proven track record in delivering on its promises. It knew sales of DSS would be hot, with one day of production worth millions of dollars in sales revenue. This operation had to be ready on time and, once operational, experience minimal downtime.

After evaluating possible vendors, Thomson chose Universal Instruments, a supplier of surface mount and insertion machines used to place and insert components on electronic circuit boards and move the boards through manufacturing.

Service and support are as important to Thomson as the integration and machines themselves. Universal has an after-sales and logistics operation that delivers spare parts and service quickly and the company tailored its on-site support to meet Thomson's specific cultural and functional requirements.

Thomson found Universal's Surface Mount Technology (SMT) Laboratory to be another benefit of this partnership. Since DSS was Thomson America's first design requiring high-volume mass reflow soldering, it relied on the expertise of Universal's SMT Lab in Binghamton, New York, to ensure successful implementation of the new technology.

To date, Thomson has exceeded all its production volume and quality goals. Since DSS was the first product of its kind for Thomson, it didn't have a standard for comparison, but the fully integrated solution significantly reduced work in process and manual board handling. Lines of communication on the factory floor got shorter and, as a result, quality issues get fixed much faster.[9]

➤ Sometimes It Isn't the Initial Cost but the Upkeep

For more than 40 years, the name Genie has been synonymous with garage door openers. In 1954, the Alliance Manufacturing Company (now the Genie Company) introduced the first mass-produced garage door opener system, making Genie the first name in the industry. Since then, Genie systems have become the industry leader based on reliability, durability, and safety. Today, the Genie Company is a subsidiary of Overhead Door Corporation of Dallas.

In addition to the well-known garage door opener systems, the Genie Company, headquartered in Alliance, Ohio, produces wet/dry shop vacuums and floor care systems, power blowers, and gate openers. Genie operates manufacturing facilities in Alliance and Baltic, Ohio, and Shenandoah, Virginia.

At the Shenandoah site, Genie manufactures the electronics for more than one million garage door opener systems each year. Since many systems require multiple boards, this operation produces more than ten million circuit boards annually. To achieve this kind of output, Genie's plant in Shenandoah keeps 300 employees busy 24 hours a day. Like Thomson, Genie relies on Universal machines.

During recent years, as demand for Genie's product increased and production goals climbed, production took precedence over preventive maintenance. However, without preventive maintenance, Genie soon found it was not getting maximum production from its equipment.

The firm didn't have a preventive maintenance program. It couldn't plan or schedule maintenance, so its equipment was breaking down at the worst possible time. Instead of trying to sell the company new equipment, however, Universal suggested a disciplined maintenance program it calls the UPTIME 100 productivity program. Its objective is to help customers improve their manufacturing operations. This is accomplished not only by addressing equipment and process concerns, but cultural and environmental issues as well. While the process varies depending upon each customer's needs, it begins with a thorough analysis of the customer's situation.

After performing the in-depth customer analysis, Universal's field engineer, Tom Schubert, suggested immediate changes to improve the performance of Genie's machines. He then instituted a program that created long-term process and cultural changes affecting everyone in the plant. Some of the alterations Schubert suggested included keeping logbooks on each machine so issues can be communicated and tracked, trends can be identified, and preventive maintenance can be scheduled; instituting team meetings so staff across all three shifts are kept informed of machine status, production plans, and maintenance schedules; establishing consistent maintenance; and developing a system to manage and track spare parts usage and costs. Additionally, Genie made changes in its maintenance staff and improved operator training and staff communication.

After the program was implemented, Genie's defect rate dropped to less than 1 percent, machine utilization increased by 45 percent, and costs for spare parts and service dropped by more than 50 percent.

The improvements realized through UPTIME 100 enabled Genie to introduce a more efficient manufacturing technique: a work-flow system that streamlines the manufacturing process. Genie's flow line emphasizes reduced work in process and on-hand inventory, while increasing board quality. It is a move away from its traditional piecework method, and it is being embraced by employees and other companies in the Overhead Door family.[10]

➤ Implementing a Fast-Prototype Facility

Solectron is a leader in contract manufacturing of electronic equipment, an industry that has grown from back-alley garages to the forefront of high tech in the last decade. The Solectron Technology Center is focused on rapid prototype production of electronic assemblies for a wide variety of companies. Rapid prototype manufacturing is critical for quickly developing and introducing products, and Solectron uses high-volume production equipment so that successful prototypes can be quickly moved into volume manufacturing.

Solectron needed a fully integrated system to meet the quick-turn manufacturing cycle, plus a simple interface to popular software used by its staff and customers. The enterprise resource planning (ERP) system needed to be integrated with processes and systems used to process the data customers need to describe their products: CAD, bill of materials, and part information.

Solectron selected business partners based on the compatibility of the company culture and technical direction. Baan was selected as the ERP vendor, with its Baan IV software providing inventory and manufacturing functions. The finance functions were integrated with Solectron's existing MRP II system, with an expected move to Baan for production units of the plant.

Solectron has many sites in the United States and overseas, and the firm uses the Internet as the backbone of its intranet. All of the Baan program data and its business process documents are shared on a Web site and can be accessed from the other sites. Customers send technical information for prototype and production to Solectron via the Internet, reducing communications from days to hours. This is critical in reducing the total cycle from two weeks to two days. Customers can access production status by dialing into Solectron.

The hardware used is Compaq ProLiant 5000 computers with four processors with 512 megabytes of memory and 16 gigabytes of RAID disk. The database is SQL Server in an NT environment. Using PC hardware and software reduced the cost to about half what it would have been with UNIX or other technology.

By implementing Baan in Microsoft SQL Server, Solectron has a quick response system that interfaces easily with other applications. The multitasking Windows 95 clients and GUI capabilities provide it with an easy and fast way to obtain and process information.

The Solectron Technology Center has achieved its objectives and is at full capacity for short-turn prototype assembly production. Baan software improves Solectron's ability to find and order parts for quick-turn production. Most of the business is now high-premium, two-day production and the prototype business is obviously a conduit for production business. About 80 percent of Solectron customers now use this capability.

The entire process is now integrated and data need only be entered once. In many cases the data come directly from Solectron customers and suppliers and are not entered by humans but by programs. Customers send orders directly using EDI or paperless fax and receive acknowledgments the same way. Cycle time has been reduced but, more important, fewer errors are introduced.

Employees can easily use tools such as Excel for their analysis and process improvements. In the past, they had to print and enter data; now they select, cut, paste, and send. These tools offer significant improvement in the capability to serve customers.

Solectron was able to improve its rapid prototype manufacturing capability to reduce the time it took to get customers' products into volume production. The total time taken to set up the operation was about four months.[11]

➤ "ROI" Spells Reengineering at KLA Instruments

KLA Instruments is the world's leading manufacturer of process control and yield management equipment for the semiconductor industry. It has 75 percent of the U.S., Japanese, Asia-Pacific, and European markets. Its systems and advanced process control techniques play a critical role in integrated circuit manufacturing by monitoring wafers after each process step, letting the manufacturer detect, analyze, and eliminate process-induced defects and improve yields.

Since its founding in 1976, KLA has steadily grown to more than 2,200 employees in the United States, the United Kingdom, Japan, Israel, France, Germany, Italy, Switzerland, Korea, Malaysia, Singapore and Taiwan. During recent years, its growth has been especially rapid, doubling revenue each year and adding more than 100 employees every quarter.

"We needed to overhaul our business processes, and with them our computer hardware and software, to keep up with such a rapid growth rate," said KLA's chief information officer, Gil McInnes. "As a key first step, we sought out low-risk partnerships with companies that were leaders in their own industries—Oracle, Sun and KPMG."

To stay on track, KLA needed to reengineer its business processes and update its computer network. When McInnes joined KLA, the company's existing systems were running a ten-year-old application with limited functionality, and they were beginning to break down. The systems—a network of IBM RS/6000 workstations—did not give KLA enough information to run the company properly or process information to help KLA grow its business.

Under KPMG's guidance, KLA mapped out 14 business processes across two of its major divisions that it wanted to reengineer immediately, and decided on a name for its undertaking: ROI, for "Radical Operational Improvement." It established metrics in finance, manufacturing, and sales and distribution against which it would measure its success. KLA also set up an aggressive time frame in which to implement ROI and convert its business processes.

Senior management at KLA took responsibility for driving process improvements, such as reducing the number of manual reports expected by sales groups, improving the accuracy of manufacturing reports, and designing a standard general ledger for worldwide use. McInnes's MIS department had to make sure the computer systems it chose would support such radical changes.

KLA turned to the Sun/Oracle Applications Technology Center (SOATC) during part of the project. It provided benchmarking and Oracle application expertise. After running its own data there, KLA was able to verify the configuration it wanted to buy from Sun.

A SPARCcenter 2000E now runs all the Oracle software for KLA's ROI program, including the Oracle Financials payables, receivables, and fixed assets modules and the Oracle Manufacturing inventory, purchase order, bill of material, work in process, and manufacturing resource planning modules. Two SPARCserver 1000Es with attached SPARCstorage Array 112s house the company's databases and other transaction information.

Because Sun and Oracle are so well integrated, and because KPMG had extensive Oracle expertise, the implementation process at KLA went smoothly. The project came in under budget and within the ten-month goal KLA had targeted.

Soon after implementing ROI, KLA saw results. Sales reporting and distribution processes are being used as they were designed to with ROI. In manufacturing, users have access to standard, up-to-date reports that are more user-friendly and flexible, and most manual reports are now on-line. For the finance department, automation has greatly increased the accuracy of reports, with some major functions being completed in half the time they used to take.[12]

■ IMPROVING PLANT LAYOUT

One interesting application of technology is in designing facilities; the same basic CAD software used to design products can also help create an efficient plant. Toyota Motor Manufacturing Canada (TMMC) is a division of Toyota Motor Corporation, one of the world's largest companies and the third-largest carmaker.

The strong yen has made exporting cars from Japan expensive, so Toyota has shifted much of its production overseas. As a Toyota car assembly plant, TMMC builds the four-door Corolla for markets in Canada and the northeastern United States. The three-million-square-foot facility recently doubled production to 200,000 automobiles per year.

Because a typical automotive assembly plant must retool every five years on average, CAD software has proven to be essential in streamlining this process. Before TMMC used CAD, hand drawings were sent over from Japan with retooling specifications, then implemented using manual drafting methods—an extremely slow, inefficient, and costly process.

Several years ago, TMMC's Facility Engineering Department, responsible for plant layout, modification, and assembly-line machinery tooling, began using Autodesk's AutoCAD design and drafting software as a tool for initial plant setup, equipment, and building layouts. Currently, engineers use AutoCAD running on Pentium-based PCs over an IBM Token Ring network, with plans to migrate to Windows NT.

The initial efficiency gains realized with AutoCAD were dramatic, and newer versions of the software increased efficiency even more. Among the improvements are easy integration with Microsoft Word and Excel, so engineers can create accurate, high-quality presentations. Toyota designers can also send AutoCAD files all over the world with ease.

Using the common database in AutoCAD, other departments at Toyota—including paint, body welding, and stamping—easily combine their information with that from facility engineering. Future plans include the installation of a digital picture and camera system that would link design files to a preventive maintenance database for use by factory service engineers.[13]

■ IMPROVING NONMANUFACTURING OPERATIONS

Not all companies manufacture products, of course. They can still use technology to improve their operations. For some firms, access to information *is* their business. Fireman's Fund Insurance Company entrusts almost a terabyte of legal documentation to a high-performance document management system powered by Compaq hardware and an Oracle database. With fully customized software to handle scanning of, identifying of, and intelligent searching for documents, an enormous amount of raw data becomes manageable, useful information.

Fireman's Fund Insurance, one of the largest insurance companies in the world, recently found itself facing an avalanche of paper surrounding environmental litigation proceedings—almost nine million pages of reports, memos, correspondence, meeting notes, and other documents of all shapes and sizes. Making good use of this documentation was a "good news/bad news" scenario for Fireman's Fund. As Ron Ryba, the company's vice president of environmental claims, explains, "Effectively managing litigation requires excellent information. We had access to plenty of raw data—millions of pages of documentation. However, we needed an efficient, cost-effective way to capture and convert that data into meaningful information that we could search repeatedly and quickly."

Fireman's Fund needed a top-notch document management system and, with millions of pages to work with, it turned to industry leaders Compaq, Oracle, Microsoft, and Executech to provide the performance, reliability, and scalability the company required.

Fireman's Fund had plenty of experience managing the legal discovery process. From using microfilm to creating an in-house system, the company had employed a variety of methods for handling data, with varying success—and costs. "Fireman's Fund is a forward-thinking company, looking for efficiencies in everything we do. Searching through millions of documents using a word processor or microfilm was obviously not the most efficient and productive way to handle this situation," reflects Ryba.

John Murphy, Fireman's Fund information systems (IS) manager, agreed and began searching for a solution that would cost-effectively give Ryba's team the power and control to manage the ever increasing paperwork. Murphy notes that "proprietary systems were certainly available, but we wanted the assurance of long-term flexibility. A key requirement of this system was that it be based on open standards for both hardware and software."

Fireman's Fund called on Executech, a company that specializes in electronic document management software and hardware integration. Executech knew that this massive project would require significant computing horsepower, tight integration among processes and technologies, and the most reliable hardware/software combination. Its recommendation included Compaq ProLiant hardware and Oracle software, which together provide the best performance and reliability at the right price.

While defining the system and its requirements, Executech and Fireman's Fund agreed that "if you're going to do a project of this magnitude, it has to be on the safest and most secure platform available. Our recommendation was Compaq ProLiant," says Ted Davis, president of Executech.

Total cost of ownership was another consideration. "We initially considered using another vendor's RAID system, which at first seemed to be less expensive. But it failed miserably at even basic read/write tests, and we realized that a pure Compaq solution was the way to go. Compaq stands behind its solutions 100%, and we've found tremendous price/performance advantages every step of the way. Compaq was already an authorized vendor at Fireman's Fund—Compaq ProLiant was the obvious choice," reports Davis.

For software, Oracle provides the database power for Executech's E-TECH software. Executech also recommended Microsoft Windows NT, in large part to take advantage of Verity's Topic Enterprise Server client/server full-text search engine, which adds additional power to E-TECH searches. This combination of Oracle and Windows NT provided yet another reason to implement Compaq hardware: Compaq ProLiant is an official reference platform for Oracle7 on Windows NT.

"Compaq's relationships with Oracle and Microsoft go a long way toward helping us implement reliable solutions for our customers," says Executech's Davis. "The tight integration of Microsoft's operating system and Oracle's database software with Compaq hardware is a great comfort factor. One tangible example of that integration is the Compaq SmartStart software that ships with all of its servers. SmartStart is a terrific tool for quickly installing and optimizing both the operating system and the database. That lets us concentrate on the business of creating the best document management solution around."

Davis likens the Fireman's Fund implementation to the building of an airplane in midflight. "We immediately dispatched some 85 people nationwide, with Compaq ProSignia file servers, scanners, and external disk drives in hand. During a ten-month period, these crews set up mini-networks in hotel rooms, scanned and indexed hundreds of thousands of pages every week onto the external drives, then shipped the drives to Executech's Connecticut office, where the system was being developed. Meanwhile, the Fireman's Fund legal team was calling into the system from the West Coast to search documents as they were added. The system was literally running as it was being built."

More than eight million pages of documents have been scanned into some 900 gigabytes of storage space controlled by four Compaq ProLiant 4500 and 5000 servers.

Fireman's Fund clearly believes the goals for its document management system have been met. "We put about $800,000 into the document management and search system. That sounds like a lot of money until you consider that we probably saved $20 million over doing the same sort of caseload on microfilm—and we were more productive along the way," explains Ryba.[14]

■ TECHNOLOGY IN OPERATIONS AND MANUFACTURING

Technology is helping to improve productivity in manufacturing at most companies, but it's primarily improving processes rather than automation. The reason is simple. Most automation battles have been won, so companies are now reengineering their organizations to operate more effectively. In many cases, this has led to sourcing out services and operations that were once considered vital—including manufacturing itself.

At the same time, companies continue to improve their manufacturing processes, including adding new types of machines and even totally new technologies. Because of the shrinking product cycles in every industry, being able to produce prototypes quickly then shift them to volume production has become a major part of the total production process. Not all operations involve manufacturing products, and technology is also assisting the many firms whose "products" are services.

3

From Purchasing to Partners

■ INTRODUCTION

No activity in a company better illustrates the trend toward overall integration of functions than what is increasingly being called the "supply chain": the continuum that begins with vendors of parts and raw materials, extends through manufacturing, warehousing, transportation, distribution, and out to customers. In optimum form, in fact, it extends back to a supplier's suppliers and forward to a customer's customers.

Most companies still treat each of these entities as an isolated function with its own requirements, but smart corporations seek to tie all of these pieces together into one *integrated enterprise*. And it's an enterprise driven by customer demand, not vendor supply.

One example of today's lean companies is Wal-Mart, which uses supply chain management to buy directly from manufacturers and sell to customers while maintaining minimum inventory. It uses just 10 percent of its store space for inventory—far less than the traditional retailer that doesn't have as much knowledge and control of its customer demand.

Perhaps no function has really been more affected by modern integrated enterprise software than procurement, which is expanding to reach all parts of the modern corporation, from sales to its traditional focus, vendors. Many modern corporations don't make products in the traditional sense; they assemble or even buy them.

In the process, at many companies, procurement has become virtual manufacturing, with all the implications that suggests about its role. This has literally transformed many companies.

■ "BUILD AND SELL" MEETS ITS DEMISE

Among manufacturers, the traditional "build and sell" model is becoming obsolete, to be succeeded by the "sell and build" company of today.

Few manufacturing companies have reached this ultimate goal: to build and deliver a product to a customer's demand, simultaneously ordering from suppliers the parts to replace it. Many companies are approaching the ideal, however. One example of the build-to-order model is personal computer companies such as Dell Computer and Gateway 2000. These firms take an order from a customer using an 800 number, then they assemble a computer with exactly the options and software customers specify, from pieces supplied by others or sometimes from parts built to order for Dell or Gateway. The companies then express-ship the computer to the customer, sometimes within hours.

Two of the biggest successes in the computer business, Dell and Gateway have virtually eliminated finished-goods inventory and traditional distributor/retail markup, allowing them to supply quality products at lower costs. They can also deliver new products quickly since they have no inventory of older components and computers to consider. Some companies go even further, maintaining no inventory at all, assembling no products, simply drop-shipping products directly from suppliers, yet still maintaining unique identities and delivering products quickly.

Such big suppliers as Compaq and Hewlett-Packard and even discount chains like CompUSA have all responded to the threat from direct, build-to-order suppliers by developing similar programs. Coming from different business environments, however, they are caught between existing channels and suppliers, who view the moves as disloyal. Nevertheless, customers clearly like the model.

It is interesting that big corporations prefer direct-purchase systems as much as penny-pinching consumers do. If anything, these firms are less likely to need the comfort of seeing products first at a local store or having local service facilities, since ordering equipment this way is more traditional to corporations than going to a storefront location anyway. And new service techniques to address such situations as replacing modules or simply getting an upgraded computer overnight remove much of the worry about buying products from a distant supplier.

Obviously, Dell doesn't order one new disk drive for inventory every time it sells a system, but it does track shipments to enable it to stay lean and yet handle customer demand. Dell and other companies—even those with more traditional sales and distribution channels—are installing sophisticated software called, among other terms, *distribution requirements planning,* which allows these companies to manage this complex process of continual replenishment. And, as industry trends indicate, it's becoming increasingly difficult to separate different functions in a corporation from one another. If the delivery of a certain product to a customer automatically triggers the addition of a replacement to the next purchase order, how can the delivery simply be considered part of a transportation logistics system?

As companies increasingly adopt systems that tie sales orders to the factory floor, they're also using them in equal measure to manage procurement. The same delivery and order also enter a data warehouse, which a marketing analyst can access to gauge product demand, perhaps encouraging the company to offer standard products with similar options in the future.

This move to integrate supply chain functions is the biggest overall trend in dealing with vendors and service suppliers, but steps that help implement this integration are equally exciting. Over the past 30 years, many corporations have been turning to automatic exchange of order, status, and billing information to reduce costs and save time, and this electronic data interchange (EDI) has had substantial impact, particularly for larger companies.

Now EDI—and other electronic commerce schemes—are rushing toward the Internet as the ideal conduit for transactions as soon as issues of standards and security are solved.

■ THE SUPPLY CHAIN: CREATING THE VIRTUAL ENTERPRISE

As in every other phase of corporate life, technology is helping to transform not only the way a company operates but even what it is.

As corporations reengineer, regroup, and reform, they're becoming virtual enterprises—chains of buyers, suppliers, and customers, all united to compete effectively in an ever more competitive market. In a sense, these new organizations are the logical successor to the vertically integrated corporation of the 1960's, which intended to compete by controlling every step in its product chain, from raw materials to retail sales.

The classic example of vertical integration was yesterday's automobile supplier, with Texas Instruments (TI) the leading advocate among high-tech companies. This model collapsed as surely as big government, a victim of costs of capital, the hopelessness of managing such a diverse enterprise, and competition from more nimble participants.

Now *virtual integration,* with new attitudes and better communication and control, allows diverse companies to work together as one, even working effectively with suppliers and customers of competitors. And few corporations have embraced the concept more wholeheartedly than those same automobile suppliers, which have shed peripheral operations and turned to outsourcing as a way to compete effectively with Japanese companies that mastered the techniques long ago. TI has also sold off its unrelated businesses, focusing on its core competency in specialized, high-value semi-conductors.

Although companies are reengineering their operations for greater efficiency, old concepts don't die. What was once called "buying," then "purchasing" and "procurement," is still with us, albeit as part of a more sophisticated process. This function mutated into MRP—materials requirement planning—which evolved into ERP, or enterprise resource planning, a term also likely to disappear as corporations and vendors overcome the significant challenge of tying all functions of a company's life into one integrated enterprise management system. Whatever it is called, however, managing the purchase of parts and materials that typically represent more than half of a company's costs is an important task.

➤ The Rise of the Supply Chain

In one sense, supply chains are America's answer to the Japanese *keiretsu,* interlocking companies that work together with common goals, serving different functions but each contributing a part to the final result. A well-known example is the one created by Apple, Motorola, and IBM, which joined together to develop microprocessor and software products to compete with the Intel-Microsoft juggernaut. A bad example, perhaps, for it has been spectacularly unsuccessful in its primary aim, but the combination did accomplish many goals, including developing a generally acknowledged better, if irrelevantly so, competitor to Intel's primary Pentium offerings.

Most supply chains are less noted, more modest in goals, and more successful. They typically represent better internal management of product development, manufacturing, and delivery, partly as a result of better partnering with vendors and customers.

The concept of the supply chain grew from many needs within corporations. The fundamental driving force was the rise of the customer. Whereas companies once were the rulers of our economy, developing products they built to inventory, then shipping to wholesalers and thence to retailers, now the customer is king. Companies need to give customers what they want, and that means they have to be nimble as well as fast.

There are four primary goals of integrated procurement and supply chain management, to give it the full description:

➤ Shorten product development time.

➤ Lower operating costs and improve return on investment.

➤ Shorten the time it takes to deliver a product to the customer.

➤ Respond quickly to changes in demand.

This view is from the top down, as management attempts to improve a company's performance. Increasing numbers of executives and even boards of directors are realizing that issues of supply are as strategic to their companies as product development and marketing, even if their solution is to turn to outsourcing rather than traditional captive manufacturing models.

➤ Integrating the Supply Chain

The secret to improving supply chain management is to discard the traditional way of regarding a company as a series of sequential functions and, instead, look at its operations as a process. And it's not a process that begins with product design and ends in a warehouse. It's a process driven by customer demand and ending with suppliers, not the other way around.

Obviously, few companies really think this way. To many managers, "supply chain management," like "reengineering," has become a code word for reducing head count. Instead of rethinking and reorganizing the process, they've simply lopped off functions, shifting some to suppliers and other companies.

One reason for this is that truly integrating the supply chain is very difficult. Few companies really have accomplished it internally, much less with external suppliers. And though software vendors promise supply chain solutions, they really only offer pieces that address certain parts of the puzzle or grand schemes that still require massive customization to fit a company's requirements. Indeed, according to Advanced Manufacturing Research, a consulting firm that specializes in this field, supply chain applications are still largely custom designed, whether for stand-alone or enterprise use. The biggest categories in this mix are products for planning, forecasting, and scheduling, plus data warehouse applications that allow corporations to better utilize customer and transaction data for analysis, marketing, and forecasting.

The difficulty of integrating distinct functions within a company, much less suppliers and customers, is heightened by nontechnical requirements as much as the need to deal with very complex operational and communications issues. For a multicompany supply chain to operate, the companies have to become true partners, ending the traditional adversarial role between supplier and buyer. This requires a level of trust and disclosure foreign to traditional American business practices, including honestly sharing cost data without the buyer taking advantage of this trust. It also requires that partners be true equals, if not in size, at least in competence and commitment.

Beyond these issues, however, successful supply chain management requires competent systems to communicate, monitor, and control the process. Supply chain management obviously must support internal process, but it also must allow the sharing of data with others. This includes not only pricing, orders, acknowledgment, shipping notices, and invoices, but product and design information, including computer-aided design software and files. Many corporations are becoming so heavily dependent on these systems that they require suppliers to use specified software.

➤ Supply Chain Thinking at Auto Firms

Automobile companies are leaders in supply chain management. Long ago they recognized that they needed a continuous chain from suppliers, so they forced suppliers to adopt and use the 3-D CAD/CAM files that have become the standard definition of their components and design. These companies no longer even use the paper blueprints long associated with mechanical design. The electronic representation is the standard.

Using computers allows car companies to quickly develop and change designs over networks, saving an estimated $1 million per day of development time. Unfortunately for suppliers to Detroit, however, the big three use different CAD standards. Ford uses a system from Structural Dynamics Research Corporation (SDRC) and General Motors uses EDS Unigraphics. Fortunately Chrysler, most European companies, and many Japanese firms—as well as GM's progressive Saturn Division—use the same software: Dassault Systems' CATIA. The industry is working toward standard interfaces that should reduce this headache for suppliers and let companies make better use of intercompany communication.

➤ Replacing Forecasts with Actuals

Along with this need to share information is a move from working with sales forecasts to instead using actual sales numbers. This again requires companies to collect information regarding customer sales and deliveries that is often difficult to obtain from distributors and retailers.

An important element in shortening the supply chain is improving logistics once a product is produced. The goal is to reduce inventory and warehousing without harming customer deliveries. An important step in this process is to replace products with information, for there's no need to maintain stacks of goods in many locations if the right part can be supplied just as quickly by building to order or shipping overnight from a central site. The revolution in communications and express shipment has allowed many companies to outsource more and more of these functions, cutting fixed costs while improving customer service and support.

A final but important issue in improving the supply chain is figuring out a way to measure this improvement. Most companies don't have adequate means to track and monitor their own operations, much less those extending away into layers of suppliers and customers. Better systems will help this process. A trade group, the Supply Chain Council, has been established by users of and vendors to this field to help set standards and resolve many remaining issues.

➤ Global Supply Chain Management

Supply chain management is such a new concept that vendors are only now offering tools to accomplish it. It's worth looking at what it involves, because trying to implement a system is a serious undertaking that calls for major commitment of time, personnel, and money. It's also sure to disrupt company operations as staff are forced to change the way they think and act, probably wondering if the step is a precursor to the elimination of jobs.

If supply chain management is particularly complex for a multisite, multinational operation with many sales and distribution channels, it also provides multiple benefits. It's difficult for such an organization to accurately manufacture and deliver products, much less manage just-in-time inventories held by suppliers and use customer demand to control the whole production chain.

The ideal supply chain manager integrates the separate functions traditionally called *distribution requirement plans* (DRPs), *master production schedules* (MPSs), and *materials requirement planning* (MRP), plus the newer supply chain plans. These integrated plans allow a company to track and manage inventory and purchasing, both in its own manufacturing facilities and warehouses and in those of its vendors and customers.

Such a system can also be broken up into more manageable modules to understand the situation in any individual site. The software will tie any desired component or materials to products that are in inventory or have been shipped, allowing automatic replenishment at certain trigger levels. It also tracks any components that could be critical, allowing manual increases in inventory level. Modern packages are graphical and easy to use, and allow users to choose a variety of views and priorities.

Clearly, maintaining minimum but adequate inventories at optimum cost is the major function of a modern procurement process. The traditional purchasing process requires careful attention by humans, even if they have access to spreadsheets and similar tools as aids. In practice, buyers and suppliers constantly seesaw, trying to find the optimum inventory level for individual components and materials. Though tied to forecasts, the source and content of these forecasts were always suspect, having little relationship to actual customer usage. Even fast-reaction schemes like E-mail and EDI couldn't accomplish much when buyers didn't know what they really needed to know.

With modern systems, there's a consistent trail from a customer—even from a customer's customers—back through an organization to suppliers. Automatic limits trigger buying without the need for individual purchase orders, and built-in alarms signal concerns such as unanticipated demand or product accumulation.

Communications ties the supply chain together. Though EDI has been the standard, it's clear that the Internet—probably using EDI protocols and standards—is the future conduit for this process. Using conventional Web browsers, companies can create intranets and extranets that allow participants to access information as easily as finding a stock price and initiate orders as easily as ordering flowers on the Web.

Supply Chain Functions

The variety of functions tracked by such a scheme is substantial. Here are the modules that make up Oracle Applications Supply Chain Management, for example:

Inventory: Lets user define warehouse structures and controls needed across multiple inventory locations. Provides the flexibility to control items by lot, serial, and revision numbers. Maintains inventory accuracy with ABC analysis and strict cycle-count calendars. Allows automated data collection devices to capture all material transactions.

Master Scheduling: Helps materials managers, planners, and schedulers more easily plan production, purchasing, and interorganization replenishments. These features are designed to help companies using classical planning meth-

(continued)

ods transition smoothly toward using on-line, interactive planning, and decision support tools.

Order Entry: Provides streamlined efficiency and customer service functions. Each sales channel can establish its own service policies to process orders. For example, users can establish channel- and customer-based pricing, credit approvals, and delivery policies. Selling locations can check product availability, allocate from on-hand or future supply, and promise delivery from anywhere in the world.

Payables: Increases productivity to process and pay more invoices with less staff, improves controls to pay only for those goods and services ordered and received, and improves cash management by maximizing discounts and preventing duplicate payments.

Product Configurator: Verifies product configurations, automatically selects configuration options, and generates manufacturing bills of material according to product configuration constraints. Helps improve order-taking and fulfillment productivity by eliminating errors in new sales orders and manufacturing bills of material.

Purchasing: Simplifies routine transactions, reduces paper handling, and provides an electronic communications framework for daily procurement activities.

Quality: Tracks and analyzes product quality throughout the enterprise and evaluates the resulting impact on customers.

Receivables: Provides full-function accounts receivable functions. Based on rules defined by the user, Oracle Receivables automatically accounts for all cash receipts, including miscellaneous receipts such as refunds and investment income. On-line collection screens help track, monitor, and collect receivables to reduce delinquent accounts. Allows control over all transactions, generating standard invoices, credit memos, recurring invoices, and commitments with minimal data input.

Service: Provides functions for postsales service management. Allows user to track the entire installed base, manage

(continued)

service contracts, record service requests, accept returns, and perform repairs.

Supplier Scheduling: Streamlines and automates the order management process between a company and its supplier partners. Frees purchasing professionals to focus on strategic supply-base management programs and those tasks that have a direct impact on the bottom line.

Supply Chain Planning: Provides a planning model that addresses the entire supply chain, providing a view of the global enterprise in which all locations, including those of customers and suppliers, are included and are part of supply chain bills of materials.

Web Customers: Customers, business partners, and internal business units can process and access business transactions according to a business-process flow that includes customer-oriented data and enterprise-wide business processes.

Web Suppliers: Suppliers, partners, and internal business units can process and access business transactions according to a business-process flow that includes all supplier-oriented data and enterprise-wide business processes.

Other suppliers offer similar products. Most, like Oracle, can integrate these supply chain modules with overall enterprise management software such as that offered by Oracle, SAP, PeopleSoft, and Baan.[1]

➤ Reengineering a Pharmaceutical Company

One company that has been through a complete design of its supply process to address the whole supply chain is Merck & Company, a major supplier of pharmaceuticals. Merck faces a problem common to its industry: On average it takes 15 years and $359 million to bring each prescription medicine to market in America, according to the *Drug Information Journal* and government economists. Only one out of the 10,000 compounds studied is ultimately approved by the U.S. Food and Drug Administration.

By redesigning its program to purchase goods and services, Merck was able to save over $400 million between 1994 and 1996, while getting goods more efficiently, reliably, and quickly.

Merck began to reengineer its procurement processes in 1992. The project sought cost savings and higher productivity while eliminating redundancy and duplication and developing processes that supported the effort, including training, technological enhancements, and a clearly defined vision. In a first step, a project team evaluated existing methods for procuring goods and services, developed a streamlined process, implemented new systems, and redesigned the work performed by procurement professionals. The team had three main objectives: Achieve customer satisfaction, improve the efficiency and effectiveness of the acquisition process, and decrease the total cost of goods and services acquired. The team defined its customers as both procurement professionals and Merck employees who purchase goods and services.

The team set three goals: Reduce suppliers from 40,000 to 10,000, consolidate product purchases for better prices, and dramatically increase the percentage of goods and services bought under contract from 20 percent to 80 percent.

The procurement team then focused on the negotiation strategies for major strategic purchases, looking at the total cost of ownership. From supplies to capital equipment, the goal was to save money while maintaining or improving quality, service, and customer satisfaction. Merck took advantage of new technologies to allow people to move from routine transactions to more strategic and valuable tasks, a shift achieved without loss of Merck jobs.

Employees can now order goods and services directly from electronic supplier catalogs, including verifying their authorizations and sending orders to the suppliers electronically. They also can check order status and spending on their computers.

To integrate the process with other critical company transactional systems, Merck created a transaction messaging system, choosing the procurement module of SAP's R/3 enterprise software suite as the transaction system to process requisitions, purchase orders, receipts, and key supplier information. R/3 was selected based on its technical architecture flexibility and scalability, coupled with its rich functionality capabilities. Since the project's geographic scope was worldwide, the supplier's global support capabilities were very important. SAP held a strong presence in all major Merck markets around the world.

The new procurement system also automated the approval process, eliminating most manual and paper activities, while significantly reducing the time required to process an order.

With this automation, purchasing personnel avoid most paper processing including recreating order information and verifying accounting and financial approval. Their role has become more strategic, with the objective of providing customer satisfaction and value, while saving the company money for medical research.

By using decision support tools and reducing the volume of transactions handled, procurement professionals were able to focus more strongly on truly valuable activities. Merck customers, procurement professionals, and suppliers now share their needs across organizational and geographic lines and expectations, trade expertise, and save money. By buying in bulk and negotiating contracts based on large volumes, Merck saved $173 million in 1996—38 percent more than the projected target. In 1995 Merck had $16.7 billion in sales revenue. For each sales dollar, the company spent $0.38 on raw materials and supplies.

The new process liberated purchasing employees from the monotonous, error-prone chore of processing 500,000 requisitions each year, with 80 percent of all orders sent to the supplier without procurement intervention. Suppliers receive orders sooner with fewer transcription errors. Electronic approval has improved the speed of processing orders without a loss of management control. Queries receive immediate responses.

The decision support system (DSS) provided the procurement department and its customers easy access to data not previously available. Rather than searching through files to find out how much was spent with a supplier or asking a computer programmer to obtain the answer from a system, the DSS enables customers to obtain procurement data from their desktops. Data access, data flow, and supporting communications create a partnership between procurement and customers for solving issues and cross-functional problems.

Prior to the introduction of the SAP R/3 system and the decision support systems, Merck sites utilized different computer systems to support their purchasing processes. In some cases, the technology was obsolete. While the SAP R/3 system was selected because of rich functionality, architecture, and vendor support capabilities, a greater technological achievement and contribution to the overall success of the project was how it was integrated into

Merck's existing information systems. Integrated into the system are data warehouses capable of providing information on current and historic purchasing activity.

In summary, using the project, Merck saved $400 million in three years, reduced the number of suppliers by 75 percent, and is heading toward having 80 percent of its purchases made under contract.

Currently 2,700 Merck people use SAP, but the benefits of the information gained from the system far exceed this number. The SAP portion of the project has been fully operational since June 1995, with additional sites added to the network through 1997. For the future, the project looks to explore new developments in electronic commerce and the Internet. This will help improve the speed and volume of catalog usage with reduced administration and overhead.[2]

➤ A Dedicated Supply Chain Solution

Fellow health care provider SmithKline Beecham Consumer Brands (SBCB) also turned to supply chain management to improve its current operations. As a leading manufacturer of more than 30 over-the-counter health products (including Ecotrin, Tums, Aqua-Fresh, and Oxy), SBCB sells to a long list of mass merchandisers, retailers, and health service organizations.

To effectively forecast demand and plan distribution and production throughout its extensive network, SmithKline has been successfully using a mainframe supply chain management system from Manugistics since 1990. More recently, the company sought to exploit the price-performance benefits offered by distributed computing. Because the software is also available for the client/server environment, migration took place on familiar ground.

SBCB adopted client/server technology because of its tremendous flexibility. It allows the company to expand the system as its needs evolve, while maintaining an integrated supply chain planning process. In today's rapidly changing computing environment, this system lets the firm become even more responsive to market needs.

SBCB will continue to use the Manugistics system's analytical forecasting and distribution planning tools to guide operational decisions. Historical sales figures, market intelligence (information that will affect the forecast), and new customer orders are factored into a statistical forecast for demand. That number is combined with marketing, sales, finance, and other departmental forecasts to create a "one-number" demand forecast that drives the distribution of products throughout the supply chain. This process has enabled SBCB to lower inventories and improve customer service.

Using a SQL relational database, the system enables planners to forecast product demand at a much lower level of detail. Instead of tracking products by brand or stocking location, the system encourages forecasts by customer, channel, or product class. By getting closer to customers' needs, SBCB can target sales promotions, respond to special customer requests, and eliminate costly misallocation of product.

SBCB says that, in addition to lowering its actual computing costs, Manugistics empowers its planners. Distributed computing brings information one step closer to the user, and the familiar graphical user interface makes it easy for the planners to access the information they need when they need it. This results in faster planning decisions, the company says.

Improved access to data also enhances planners' responsibility for their forecasts. With distributed computing, planners now can "own" product classes, and system-supported performance measures will enable management to highlight successes. In offering this capability, the new system supports the "Simply Better" program instituted by SBCB's parent company, SmithKline Beecham PLC, in which all areas of the organization seek to improve the quality of their operations. The system also flags areas that require additional attention.

SBCB sees Manugistics' integrated, demand-driven supply chain management and leading-edge client/server technology as a winning combination. By continuing to deliver what its customers demand when they demand it, SBCB can continue to provide high levels of service while eliminating the need to hold excess inventories. And with the lower computing costs, more detailed forecasts, faster planning cycles, and flexibility offered by client/server, SBCB can meet its current business goals while protecting its technology investment into the future.[3]

Another firm turned to Manugistics' supply chain management software to solve a particular problem: too many back orders.

Bridgestone/Firestone, the $5 billion tire supplier, found a solution to this challenge: time-phased replenishment planning. The firm ships 40 million tires each year to more than 6,000 customers, including 1,500 captive American Tires and Service Company stores. Its challenge was making sure that each tire is in the right place at the right time, based on forecasts of future product demand.

It used to be impossible; there were times when the company faced 300,000 back orders, a serious customer service deficiency. At that time, the company was using a 20-year-old order-point logic. It replaced this system with time-phased replenishment planning, which relies on actual customer demand, not established stock levels, to drive manufacturing and distribution plans.

At Bridgestone/Firestone, this was necessary to plan proactively for current and future orders, eliminating the reactive method of filling incoming orders as they were received. Today, planners can determine the quantities of raw materials to order, the most efficient way to plan production, and the best distribution plan to ensure on-time delivery. The result is delivery of the right products when and where the customer wants them.

To achieve this, Bridgestone/Firestone uses Manugistics, an integrated set of supply chain management applications, along with the firm's own master production scheduling system. Since the start of the project in 1989, Bridgestone/Firestone has seen marked results, including lower inventory levels, decreased transportation costs, and fewer back orders. The process has worked so well that when the company added an additional line of products in 1991 and doubled the number of stock-keeping units handled by the system, planning went without a hitch even with almost half the previous total staff. Using time-phased replenishment planning, the company provided better service to its customers—with half the inventory it had carried before.

Because supply chain management software gives planners an accurate picture of the demand pipeline, they can now identify potential bottlenecks, change plans according to capacity and warehouse constraints, and eliminate problems before they occur. The software also provides forward planning for throughput in warehouses, and the system helps ensure that personnel and space are available to meet proposed schedules.

Building on success, Bridgestone/Firestone is also using Manugistics software to forge stronger communication links with its suppliers. The company sends purchasing requirements to suppliers in Japan using electronic data interchange. Eliminating paper-based orders, this communication has made the planning process easier and faster for companies at both ends of the transaction. The effort's success has led to an initiative to create similar EDI links with suppliers in South America and Europe.[4]

➤ A High-Tech Star

As might be expected, information technology companies are some of the biggest proponents of supply chain management. Hewlett-Packard Home Products Division, headquartered in Cupertino, California, makes the Pavilion personal computer, a high-volume, product with a short shelf life, thin margins, and a technology that quickly becomes obsolete. To make the product viable, HP coordinates its efforts with suppliers in Europe and Asia that build its components, and with an assembler in the United States. HP uses Red Pepper's Enterprise Planning software from PeopleSoft to manage the suppliers, the factory, and the inventory in its retail stores. HP notes that enterprise planning is a valuable tool for processing and presenting large amounts of data, giving visibility to not only present inventory levels at retailers or the factory, but also projected inventory based on expected demand, production levels, and receipt of components and other materials. The product also lets HP run various scenarios based on different conditions, allowing the company to plan for various situations.

At the other end of HP's spectrum from consumer computers is its Test and Measurement Organization, which makes oscilloscopes, voltmeters, and other sophisticated instruments. Ten of the organization's 36 sites have installed PeopleSoft's Enterprise Planning and Production Planning Software. For the group, the main objective is shortening the time between customer order and product delivery for 20,000 products it offers. It also wants to be able to make more reliable commitments about delivery to customers, improve productivity, and carry less inventory to reduce costs.

Unlike the computer group, the Test operation has many products made in low volume. Its customers often want custom products, yet they want to pay for only what they need when they need it. The supply chain software helps the division understand its capabilities to provide these products.[5]

■ ADOPTING OBJECT TECHNOLOGY

As if writing and instituting a supply chain management system in a client/server environment weren't enough of a challenge, McKesson Corporation went one step further. It used object technology to implement its SupplyNet, a system the company created to help its customers manage improving health care quality while managing costs. It's an unusual role for a pharmaceutical firm, perhaps, but the company has found that developing turnkey information systems is a good way to build close relationships with its customers. One reason McKesson has taken this step is that the pharmaceutical business is characterized by enormous revenue streams and little profit. McKesson wanted to optimize its share of those profits.

SupplyNet, which was first tested at UniHealth, a Southern California–based health care provider, was built using object technology, which McKesson says provides the flexibility and modularity the health care market requires. Using the technology, the company can configure custom solutions for customers with basic object models. Specifically, each application is split into components that can be reused with object technology's characteristics of encapsulation, inheritance, and polymorphism. These can be used to satisfy specific needs.

At the same time, the basic application McKesson developed is already useful for many customers. The server portion of the application runs on an Oracle database, which can store objects on Sun Microsystems' SPARCstation 10 and 20 UNIX-based workstations and on SPARCserver 1000 and 2000 servers. The application is written entirely in Smalltalk, an object-oriented programming language from ParcPlace-Digitalk of Sunnyvale, California, and uses the Gentia DB multidimensional database from Planning Sciences, for decision support and on-line analytical processing (OLAP). The client side runs on Microsoft Windows 95 and Windows NT desktops communicating with the servers using an object request broker object-based message bus.

McKesson sees the application heading for intranet and Internet implementation, since it's too expensive for most users to install locally. Using a standard browser, the remote user can access a system on a standard personal computer or even a "thin" network computer.[6]

■ THE LURE OF ELECTRONIC COMMERCE

To use supply chain management, companies have to be able to communicate electronically. More specifically, they have to be able to exchange documents related to financial transactions, engaging in *electronic commerce*. Though electronic commerce is receiving much popular attention as companies increasingly conduct business on the Internet, in truth electronic commerce has been conducted for years over private networks between and within businesses.

Electronic commerce has a lot of appeals: It accelerates the time required to conduct business, shortens cycle times, reduces errors, and reduces costs. This is a direct result of eliminating human intervention, which reduces the time involved, as well as the staff cost. Processing a typical purchase order costs the typical company $5; with electronic commerce, it costs less than $0.25.

Most companies using electronic commerce want it to extend throughout their supply chain, from manufacturers to subcontractors, materials and component suppliers, wholesalers, retailers, customers, and sales organizations. Other members in many supply chains are government agencies that track safety and import regulations, even taxes. Insurance companies, banks, and transportation companies may also be involved. It's a global technology, allowing companies to transfer information and costs between different economies, whether within or outside their own family.

Among the transactions transferred by a typical electronic commerce system are purchase orders and confirmations, shipping notices, and invoices. The same system can also be used to exchange other types of data such as requests for quotes, price quotes, insurance claims, product information, and testing specifications.

One advantage of electronic commerce is that it provides a complete record of events. Like Federal Express, the system always knows the status of transactions.

The specific mechanism used for most electronic commerce is electronic data interchange. EDI lets companies and their partners exchange structured data between computers. It became widely used in the 1970s, initially in retail, transportation, and medical businesses. Now it is widely used in manufacturing as well, with about 75,000 regular users.

EDI can be implemented in many ways, but typical users install suitable software available for any type of modern computer, from PCs to mainframes. They then exchange information over an electronic clearinghouse or value-added network (VAN). Obviously, the Internet has the potential to replace the VAN, although there are security and technical issues to be addressed and solved before this is widespread. At present, small users can use a service bureau to handle its transactions, typically communicating with the bureau via fax.

➤ EDI at Compaq

One of the biggest fans of EDI is Compaq Computer Corporation, which even promotes the process prominently on its Web home page.

Compaq began using EDI with U.S. suppliers in 1988 when it was still a new concept. Now it strongly urges both its vendors and customers to adopt EDI for exchanging data with the company. Compaq lists many advantages of EDI: improved speed and accuracy in communications between companies; lower costs for customers, suppliers, and Compaq; shorter cycle time to manufacture and deliver products to customers; better customer service; and the possibility of just-in-time production. It's also a major step toward a paperless environment for the company and its suppliers and customers.

Compaq uses EDI in many areas of operations. It exchanges catalogs and price lists and provides sales and inventory information using EDI. It uses it for orders, acknowledgments, order status, and changes, and provides shipping notification, proof of delivery, and customs information. It also supplies invoices, statements and payment applications, and remittance over EDI.

If the business partner is a customer, for example, an electronic purchase order can flow directly from its purchasing system into Compaq's order management system. Conversely, Compaq can send shipment information electronically to the customer, and the information will be deposited in the appropriate business application. All across the organization, people can know what is being shipped and when it should arrive. Once an electronic invoice is sent, accounts payable can act upon it and a signal to pay can complete the cycle.

EDI software, consistent EDI standards, and the use of data communications, usually through a third-party value-added network, make transmission possible and secure for the sender and receiver. Compaq adheres to ASC X12 standards established by the American National Standards Institute (ANSI) as well as the Electronic Data

Interchange for Administration, Commerce and Transport (EDI-FACT), established by the United Nations. The firm also participates in a variety of EDI organizations of computer-related companies that establish business models and business guidelines.[7]

■ EDI AND THE REENGINEERED COMPANY

Companies everywhere are reengineering their business processes. In many companies, Compaq included, EDI goes hand in hand with these new processes.

Using EDI, for example, Compaq has reengineered its entire procurement process so that it's faster and more accurate. Real-time data flows 24 hours a day, 7 days a week between Compaq and raw-materials suppliers. Not only is EDI used to communicate an electric "pull signal" to its suppliers when it needs more raw materials, suppliers notify Compaq when shipments are made. These notices and supplier bar coding of materials help expedite the receiving process.

Compaq customers, meanwhile, have their own reengineering projects that rely extensively on EDI. The company is looking at everything from pricing to stocking strategies to payment authorization. EDI can deliver huge cost savings—in the millions of dollars. Other benefits are accuracy and the opportunity to free up the time of professionals to plan for the future instead of day by day.

➤ No Choice in Adopting EDI

Like many other manufacturers, New Balance Athletic Shoes received a gentle push from a major retailer to use EDI—or suffer a chargeback. The company was given three months to begin electronic trading and it turned to Sterling Commerce for an EDI solution.

New Balance was up and running in three months with Sterling's GENTRAN PC product. The company was receiving purchase orders and sending invoices electronically. However, as its EDI program expanded, the company outgrew its PC package, so it turned to Sterling for a more robust UNIX EDI solution.

The firm is still receiving purchase orders and sending invoices, but it's also receiving advanced ship notices, maintaining an electronic universal product code (UPC) catalog, and using UPC bar code labeling on all products. Each of these EDI transactions is integrated into its appropriate business system. Though pushed into EDI, New Balance enjoys a number of benefits: greater partnership with its customers, better reliability and responsiveness, and elimination of communications lags with its customers. Now all shipping, invoicing, and ordering information is exchanged electronically and processing documents costs less, with significantly fewer errors.[8]

Another convert to EDI is Cummins Engine, a Fortune 500 manufacturer of industrial engines with distributors around the world. It wanted to implement electronic commerce (EC) to better coordinate the movement of raw materials, information, and products; to streamline operations; and to achieve faster, more efficient production.

To fill orders more efficiently, the company sought to synchronize the delivery of information and materials, and to integrate data about incoming shipments into planning and production operations. The manufacturer implemented numerous EDI documents such as price lists, materials releases, advance ship notices, discrepancy reports, and payment order/remittance advice transactions—all of which can be exchanged with the company's trading partners and divisions in North America and around the world. Sterling provided EDI-compliant PC software, electronic forms software, and a document turnaround feature.[9]

For companies that don't have intensive needs, Sterling developed a less expensive version of its PC-based GENTRAN: Director package for use as the need arises to trade electronically with additional large customers. It also created PC-based electronic forms software distributed to non-EDI-capable suppliers. The supplier fills in the blank electronic forms on-screen and sends the form via E-mail to COMMERCE: Network, where it is deposited into the manufacturer's electronic mailbox in EDI format.

The COMMERCE: Forms package may also be used to receive EDI documents from the hub in human-readable formats. And a unique turnaround feature enables suppliers to use the received documents to create and send related messages that reach the manufacturer as EDI transactions.

The system results in a number of benefits: shortened transit times, no rekeying, effective and accurate management of in-transit inventory, and elimination of paper invoices.

Even major telecommunications companies sometimes need a nudge to go electronic. With global operations, Northern Telecom of Mississauga, Ontario, is one of the world's most diversified developers of communications products, systems, and networks. In the mid-1980s, one of its major customers wanted to send purchase orders and receive order acknowledgments and advance shipment notices via EDI. Northern Telecom complied with an EDI implementation that was originally PC-based and served that single trading partner. This program gradually evolved into a larger, mainframe-based project with multiple trading partners.

In 1994, the company purchased a GENTRAN: Server UNIX client/server solution for its North American operations. This is Sterling's intelligent electronic commerce messaging gateway for enterprises with distributed systems, complex messaging environments, and a need for reliable untended operations. Northern Telecom is currently engaged in a global, company-wide migration to GENTRAN: Server 3.0, which facilitates Northern Telecom's global value-added supply chain operations.

EDI is an integral part of purchasing, customer orders, and accounts payable processes for the company. It has implemented EDI as part of its worldwide logistics and traffic management operations, as well as for communications with U.S. customs brokers. EDI documents are also sent internally when one division orders from another. In addition, the company uses EDI for filing tax documents.[10]

It's not just manufacturers that use EDI. Lansdale Warehouse is a full-service warehousing company located in Lansdale, Pennsylvania. One of its customers, CC Shoes, exchanges purchase orders and invoices electronically with Spiegel Catalog Company. It receives approximately 800 orders per week from Spiegel.

The orders are received electronically using EDI/EDGE with a PO/Invoice form designed for Spiegel. Lansdale prints the orders and manually keys them into its Accuplus inventory management system, which resides on an HP computer running SCO UNIX.

Accuplus will not accept orders for merchandise that is not in stock; these orders are placed aside and reentered when inventory arrives. As orders are shipped, purchase order (PO) information is transferred to the Clippership shipping program. Clippership computes the weight and shipping charges and produces a summary report, which lists all the POs, lot numbers, and weight and shipping charges. Lansdale retrieves each order in EDI/EDGE; manually enters the invoice number, ship date, shipping charges, and quality shipped; and saves it for transmission as an invoice to Spiegel.

With this solution, Lansdale is able to create a completely paperless supply chain. Orders are received, filled, shipped, and invoiced without a single piece of paper being printed. This process has eliminated several hours of data entry work each day and has resulted in reduced errors and a faster order-fill time.[11]

Safeline Corporation in Denver, Colorado, manufactures a combination car seat and stroller for infants. It receives approximately 800 orders per week from Kmart. Each order shows the quantity to be shipped to each Kmart store.

The orders are received electronically using EDI/EDGE with a PO/Invoice form designed for Kmart. In the past, as orders were shipped, Safeline retrieved each one in EDI/EDGE; entered its the invoice number, date, and quantity shipped; and saved it for transmission as an invoice to Kmart. Safeline spent eight hours per week manually entering invoice information and correcting errors.

In addition, Safeline's accounting system is not capable of handling 800 orders per week without a significant decrease in system performance. As a result, it would manually calculate the total quantity shipped and total invoice amount every week, and enter them as summary journal entries in the accounting system. This procedure required another four hours of work each week. This was a lot of wasted effort for the small company with its limited resources and limited workforce.

The company now uses EDI/EDGE with TracerX. When orders are received, the import/export function of EDI/EDGE is used to send data to TracerX, where it is placed in a database. TracerX then creates a Windows dialogue box for manually entering the shipment status for groups of orders, rather than one order at a time. This information is added to the database, and the invoice date is automatically picked up from the system and added at the same time. When an EDI communications session is started, TracerX creates a file from the database and sends it to EDI/EDGE, which translates the data into an EDI message and sends it to the value-added network. At the same time, TracerX creates a file that is sent to an Excel spreadsheet. It produces reports of all orders shipped and invoices issued, and calculates the summary journal entries for the accounting system. The new process is saving Safeline Corporation more than ten hours per week, 65 person-days a year, in data entry and error correction.[12]

For some companies, a PC isn't adequate for EDI exchange. A small Abbott manufacturing plant in Ashland, Ohio, produces rubber and plastic medical parts for pharmaceutical giant, Abbott Laboratories. This division manufactures latex gloves, IV tubing, components of drug delivery systems used in hospitals, as well as baby bottle nipples and hot water bottles.

When Abbott Laboratories implemented its EDI program, the plant reviewed both a PC-based system and use of the company mainframe system in Abbott's Chicago headquarters. The division decided to handle the accounting for the division, including manufacturing, distribution, and purchasing, using an IBM AS/400 computer with CA-PRMS software. It uses the same system, but instead chose to use it with Premenos EDI/400.

Like many companies, Abbott began its EDI project slowly. It first began receiving customer purchase orders, then added invoices back to trading partners, and is now sending purchase order acknowledgments. Next will be some shipping information and then purchase orders to vendors.

Holly Kay, senior systems analyst for Abbott Laboratories, is currently trading with two partners, one significant with larger order volumes and one much smaller. Trading, however, is steady.[13]

Shimano American Corporation, a distributor of fishing tackle, bicycle components, and automotive transmission parts and accessories, found itself in a reactive mode when it came to implementing EDI. Customer mandates were loud and clear: Become EDI-capable or face reductions in orders. This is not an unusual situation for suppliers trading with large customers, such as Wal-Mart and Kmart, which often insist on paperless trading.

In the days before EDI, Shimano American, like many companies today, manually entered orders into its JD Edwards Order Entry System. Orders were acknowledged by fax. The average turnaround time was two to five days from receipt of an order to release of the order by the order-entry system. It wasn't bad, but not the one-day turnaround time Shimano provides to its customers today.

Shimano American resisted the easy temptation to implement a PC-based EDI translator. Supporting more than 100 LAN users in the Distribution Center, Operations, Quality Control, Customer Service, and Sales and Marketing on an AS/400 D60, it knew it wanted an EDI Management System that could interface directly with its AS/400-based applications, such as JD Edwards and Varsity Shipper.

It chose Premenos's EDI/400. The benefits received more than justify the effort expended. Shimano American's orders are generally quite long, so electronic entry, in contrast to manual entry, alone has resulted in considerable savings. In addition to a major reduction in order-entry time, other benefits include a lack of errors and fewer changes of orders.

Today, Shimano American has reversed roles and is now the proactive EDI partner looking for trading partners willing to trade electronically. Shimano American currently trades with five trading partners and plans to add seven to ten new partners this year. Documents traded include invoices, planning schedule, purchase orders, shipping notices, shipping schedules, text messages, payment orders/remittance advises, lockbox, purchase order acknowledgments, and shipment information.[14]

■ THE INTERNET: THE FUTURE OF ELECTRONIC COMMERCE

All these companies have used EDI over a value-added network, but many companies are rushing to use the Internet instead. CommerceNet, an association of Internet commerce companies, is sponsoring efforts to make this widespread.

One company that didn't wait for EDI to come to the Web is Avnet. As a worldwide computer distributor, Avnet was quick to realize that the Internet could be utilized as a breakthrough way for reaching customers.

Avnet had no low-cost, low-maintenance, direct EDI solution for occasional small-order customers to call in and order parts from their warehouses. These smaller customers didn't require a dedicated EDI computer link to Avnet; their lower volume of business didn't warrant the expense. Avnet wanted a page on the Internet that would give smaller customers quick, convenient access to

Avnet's EDI system from any computer with a modem. With multiple locations all over the world, Avnet needed an immense computer network that was highly secure.

The firm hired Roswell, Georgia, systems integrator CTS to do the job. CTS knew its first step was consolidating all of Avnet's Internet servers. After conducting research to choose the best equipment and software tools for Avnet's environment, CTS recommended Hewlett-Packard 9000 servers and Oracle Database as an application tool. The HP 9000 was selected for its growth path, flexibility, and reliability. CTS provided solutions for the E-mail system, technical publications, on-line catalog, purchase order capabilities, fax capabilities, and access to shipping companies' World Wide Web (WWW) servers. Then a detailed design for a firewall security configuration was implemented to protect Avnet's internal networks. CTS implemented Checkpoint's Firewall-1 Commercial Application, and installed and configured screening routers and Hewlett-Packard servers, consisting of the K-200 Web Server, the D-350 FTP Mail Server, the C-100 Firewall Server, the C-100 Proxy Server, and the D-350 Intranet Development Web Server. It also educated Avnet staff on the new technology. At the completion of the assignment, CTS performed the ultimate security analysis for·the system. It attempted to break in to Avnet's internal network. The system remained secure.

Avnet now has a faster, higher-quality level of customer service. More information is available to more people, which has helped to increase Avnet's base of potential customers. The return on investment for this electronic commerce solution was achieved in less than 90 days.[15]

Simply exchanging information isn't the ultimate in electronic commerce for most companies. They want greater access to information and easier ordering. One way to do that is to make electronic catalogs available to their staff.

➤ Using On-line Catalogs

SAP AG offers a Web-based purchase requisition system that directly integrates with third-party catalog systems, providing a simple way for customers to buy from approved catalog suppliers.

The system provides a direct link between SAP's enterprise R/3 purchasing system and Web-based catalog systems on corporate intranets or the Internet. Employees can shop and buy using any Web browser. Users browsing a graphical catalog can select a wide variety of maintenance, repair, and operations items; the system automatically transfers orders into R/3's purchase requisition system. Companies reduce costs by receiving materials more quickly and reducing administrative costs for entering and processing orders.

These catalogs can be stored on a corporate intranet inside a company's firewall, at an authorized supplier's Web site, or at an external catalog aggregator's Web site. The system operates a "shopping basket" interface, through which goods and services are selected and a running tally of information is collected to create a requisition.

SAP developed the Web-based catalog and open interfaces with Aspect Development; ACQUION, a provider of reference data for a wide variety of maintenance items; and CADIS, a leading provider of on-line catalog software products.

SAP and Aspect also teamed up in another effort to reduce procurement costs and time to market. The firms have mated Aspect's Component and Supplier Management (CSM) system with SAP's R/3 supply chain management capabilities in design, engineering, and procurement. The integrated solution provides companies' engineering and procurement departments with a consolidated view of technical and business information for component parts and suppliers, significantly reducing design-to-manufacture time. The agreement was driven by customer demand within electronics and other discrete manufacturing industries, in which most component and supplier decisions are made by design engineers.[16]

One customer is IBM, which estimates that for every 1 percent improvement in component selection and procurement, it saves $200 million a year. Combining technical component and supplier information with business information provides a bridge between engineering and procurement departments. The joint solution enables engineering and procurement to select and purchase parts most appropriate to a design based on Aspect's technical part and supplier data on nearly 2 million parts from nearly 700 suppliers worldwide plus SAP's business information on lead time, supplier reliability, volume pricing, and logistics. In addition, design engineers can rapidly evaluate trade-offs between technical and business requirements, and can classify, manage, and reuse existing design knowledge to further shorten the design cycle.[17]

➤ Chrysler Opens Information to Suppliers over the Web

As a manufacturer of vans, sedans, Jeep vehicles, and light trucks known throughout the world, Chrysler Corporation needs to communicate with almost 20,000 suppliers of auto parts, packaging, and technology on a consistent basis. Originally, the company archived supplier data in its mainframe computer and its Supplier Communications Group mailed paper memos to inform suppliers about policy, procurement, and inventory methods. In 1993, Chrysler allowed its suppliers to access some of its mainframe systems.

Then, in 1995, Chrysler wanted to give the suppliers access to a newly implemented distributed application. The firm needed to determine how to get the applications to its suppliers, but one challenge was that it didn't know what kind of hardware and software suppliers used. Also, the company's security organization didn't want it to distribute the application code, which was proprietary and gives it a competitive edge.

Suppliers had dial-in access via PCs to Chrysler's mainframe system, but mostly they communicated using phone and faxes, and they were trading a lot of paper. It was very difficult to distribute applications even internally—there were so many PCs with various amounts of memory, several thousand users, and many different applications. When Chrysler looked at that and asked its technical staff what it would take to distribute applications, it was just about impossible. The company decided that there had to be a more effective way of communicating with suppliers than telephoning, faxing, holding countless meetings, and mailing disks back and forth.

Finally, with the advent of Internet and intranet solutions, Chrysler decided that allowing suppliers to tap into a secure intranet was the way to go. The firm has a major business objective to extend the enterprise and facilitate communications by sharing information and providing access to its systems. It seemed that a Web browser could make that possible.

There were departments at Chrysler that were already looking at Web-based access and talking about external integration via the Internet, and others were looking for a database access solution, but security was a big issue.

As Chrysler investigated an intranet solution for its suppliers, it called on one of them, IBM, to help sketch a plan of action.

Together, they sketched out a solution on a napkin in July 1995, did proof of concept in late 1995, and completed the security sign-off in March 1996.

The finished product, the Chrysler Corporation Supply Partner Information Network (SPIN), is an umbrella support environment under which other applications run. It enables communication and groupware with suppliers. Anything that Chrysler previously communicated to suppliers via paper will eventually go up on the SPIN site.

The Chrysler Corporation SPIN site is now part of Chrysler's Extended Enterprise. The site currently hosts a procurement analysis and strategy application, which generates a series of reports on suppliers and allows Chrysler's management to know the status of each supplier's business. There are plans to add two or three additional applications by 1997.

Any supplier that does business with Chrysler should have SPIN, which Chrysler will provide to a company. Ideally, any new supplier will receive information about SPIN and establish a supplier code and a SPIN ID, and then the supplier will be on.

Chrysler has about 2,300 suppliers supported now, but it does business with 20,000 suppliers per year. The first major SPIN Web application for total supplier access was supplier invoices, brought up in the first quarter of 1997.

As the Chrysler Supplier Communications Group puts more information on the system, it expects to attract more suppliers. It says cost discouraged some suppliers from signing up for prior electronic channels such as EDI. It wants to establish SPIN as a cost-effective solution for suppliers that need to sign up for Internet access only, an inexpensive approach.

The SPIN solution is part of a large number of initiatives aimed at closely integrating suppliers with Chrysler as a team. It moves them closer toward their goals of cost reduction and improved performance.

Chrysler Corporation SPIN is a three-tier platform that interfaces with Open Environment Corporation's Entera, the middleware, or glue, that generates various pieces of code. SPIN runs on a Sun Microsystems Solaris Web server inside Chrysler's firewall, which makes it easy to administer updates and controls.

One issue that Chrysler will address in the future is how to get Chrysler's design CAD/CAM coordinated with SPIN so it can host information on packaging, tooling, and design for suppliers. There is also discussion of enabling CATIA computer-aided design files via SPIN.

The solution improves existing relationships, but also can be used to develop new programs. It supports everything from developing product to delivering parts and sending payments. Chrysler Corporation SPIN is for all of its different types of product suppliers—auto parts suppliers, suppliers that do packaging of parts, and others.

Chrysler is doing all that it can to make the system user-friendly. For example, when a supplier establishes a session by logging on, the system won't ask the user to log on again to access different applications—yet security still exists. In terms of application security, suppliers establish a relationship via their user IDs.

In another program, Chrysler Corporation uses IBM/Lotus Notes to collect and manage suggestions made by suppliers to lower costs. The company claims major savings through its SCORE (Supplier COst Reduction Effort) campaign. SCORE, a program established by the automaker in late 1989, challenges suppliers in Chrysler's Extended Enterprise to continuously seek out and identify opportunities to drive costs from the vehicle manufacturing process. The program using Notes, established in 1996, has saved about $1 billion a year so far. Cumulatively, Chrysler has achieved $2.5 billion in cost savings through SCORE.

Every week Chrysler's suppliers submit more than 100 ideas, or proposals, that offer practical ways for the firm to reduce costs. Through SCORE, suppliers are encouraged to submit proposals designed to reduce costs in a variety of areas, including design, manufacturing, logistics, sourcing, and administrative transactions. What's unusual is that the company is pursuing efficiency, quality, and affordability without eroding its suppliers' profit margins. In the past, suppliers were simply ordered to reduce prices for their services and parts. Today Chrysler works with its suppliers as partners to discover ways to be more efficient and to mutually reduce costs.

Chrysler Corporation established the SCORE program in an effort to work closely with its suppliers and to identify opportunities to reduce costs from the vehicle manufacturing process. Since the program was introduced, Chrysler has entertained more than 16,000 supplier-submitted proposals.[18]

■ TECHNOLOGY IN PROCUREMENT

Technology has helped revolutionize the whole supply chain; there's no way a company could utilize a modern, just-in-time production system without advanced computer management of the processes and automatic electronic communication. Companies that once would have manufactured complete products now buy subassemblies to install at the last minute—or they might even buy whole products. This obviously makes the suppliers more important, and many corporations regard them as valued partners rather than just vendors. They find it complicated but important to manage the whole supply chain from vendors to customers, but at the most advanced companies, in fact, purchasing decisions are triggered by sales or even deliveries to end customers.

An important element of managing the supply chain is communications, and the most common solution is electronic data interchange, a standard way of exchanging information about orders and other data. It's been the most prominent type of electronic commerce until recently, but the Internet, possibly using EDI protocols, seems likely to become the standard in the near future. Some companies have even reached the state of using electronic catalogs that allow employees to order office supplies and repair and maintenance items on-line, either from external vendors or from in-house stores.

In total, today's corporations are dependent on technology, whether they make cosmetics or computers. Managing the supply chain—including manufacturing—has had enormous impact, even though it's not been completely implemented at most companies. As this management is implemented, corporations are sure to become even more efficient and productive.

Marketing to Many—
Or One

■ INTRODUCTION

As with all functions of the corporation, marketing and promotion have been vastly affected by the proliferation of technology. Many of the changes are so widespread that we don't even think of them, whereas others are just starting to have a significant impact. Technology has already transformed marketing activities such as preparing presentations and promotional materials, but will likely have the greatest impact in using customer information to conduct research and target messages. It's the catalyst that is turning the mass merchandising of the past into today's focused marketing, where narrow niches, even individual consumers, can get personalized attention. Internet aficionados are just beginning to use this strategy as they browse the Web.

The technology that makes this practical is the *data warehouse*. The concept is simple: Collect data about sales and customers, then make it easy for nontechnical users to analyze and utilize the information. Companies mine this data to find useful ideas for developing new products, targeting messages, anticipating trends, and reaping other benefits.

These systems are every employee's answer to Executive Information Systems, a hot topic of the past. "A lot of companies talk about executive information systems," says Bill Gates of Microsoft. "Although the idea behind them is probably good, the notion of having an information system that only executives look at couldn't be more inappropriate. The last thing you want is for the CEO to be looking at a system and call the people down the organization and have them looking into a different system and seeing different numbers.

"You want to invest in one system that goes down to the ultimate level of detail, and everybody can traverse the hierarchy of summarized information in a very, very flexible fashion. CEOs are the last people who need executive information systems because there are a lot of people out there helping to manipulate and find things in the data for them. It's really the organization as a whole that needs it, and needs it in a very common way."

He notes, "Five years ago, the idea of having your sales results in a pivot table, where you could just sit and see it in different ways, was unheard of. Today it's a mistake for every company not to have the information in that form as the only form. When we review product groups or country results, we bring the pivot table up onto the screen, and the person doing the presentation has highlighted the numbers that are of particular note—where they've done well, where they've done poorly. And we key off of that to have the discussion, including the ability to interactively sit there and see the information in different ways."[1]

Data warehousing has attracted the attention of corporate executives in many firms as a way to significantly improve their marketing efficiency. It's not a concept likely to garner wide public attention, however. Another new technology certainly already has.

Not since the advent of television fifty years ago and its subsequent rise to dominance has any new marketing channel had the potential that the Internet holds. Still far from its potential, the Internet could surpass television in impact, as cable, satellite, and other technologies substantially increase its speed and distribution. The Internet could reach wide audiences one day, and it's beginning to hit the narrow niches that marketers dream about. It is already having significant impact, and it could soon revolutionize such huge promotional activities as classified ads, telephone directories, catalog and other direct sales, travel and lodging reservations, and product and marketing information. The Internet hasn't yet damaged existing businesses, but they're looking over their shoulders. Many suppliers of these services are hopping onto the Internet for fear of being bypassed.

So far, few consumers seem willing to pay directly for services they receive from the Web. Likewise, advertising models borrowed from publications and broadcast remain unproven. One of the biggest impacts of the Internet, and particularly its Web incarnation, is the flattening of the information chain. Companies can now talk directly to prospects without the need for a third party, whether that third party is a newspaper reporter, a magazine publisher, a car dealer, or a travel agent. This has enormous implications for businesses that can exploit these efficiencies. But it also threatens the massive sector of our economy made of people who serve those intermediary functions.

Even as companies wrestle with fundamental changes in their distribution and customer relationships, an underlying trend is the merging of marketing, sales, and support activities at corporations. While once distinct functions, these customer-oriented functions are increasingly becoming parts of the same continuum. No longer is it possible to separate marketing from sales, or sales from customer support. They all need to follow similar principles and maintain similar cultures if a company is to have an efficient and productive interface with its prospects and customers. Smart companies now realize that their best prospects are their current customers, and they are developing and implementing systems that exploit that knowledge. It's important to involve customers and prospects in developing new products and services, but it hardly makes sense to create new systems for doing this when many companies have already collected vast amounts of unused data.

That said, the traditional role of marketing is identifying, attracting, and influencing prospective customers. Let's look at how technology has changed marketing and promotion, starting with some wide-ranging but tactical activities.

■ CREATING PROMOTIONAL MATERIAL

Traditionally, one of the important functions of marketing has been defining, managing, and producing promotional materials, beginning with proposals and presentations aimed at internal management, then progressing to sales reps, distributors, and retailers, and, ultimately, to customers. While this process often involves using specialists to produce advertising and sophisticated graphics, many marketers create their own presentations. This has many advantages, including lower cost, quicker turnaround time, and the opportunity for the expert to say exactly what he or she wants. Using PowerPoint, Freelance, or another such program, anyone can produce an attractive presentation very quickly. Whether presented on screen in a colorful imitation of television or printed inexpensively on page-size color overheads, a user can go from ideas to presentation in little more time than it would take to type the same ideas in a word processor, much less turn it over for someone else to produce.

Overhead slides and on-screen presentations are ideal for internal meetings or small groups, and woe to anyone who has an amateurish or, worse, no presentation, when a competitor does. For larger groups, traditional 35-mm slides are still popular.

It used to be a joke that consultants don't know any more than anyone else, it's just that they have everything they know on slides. That doesn't work anymore, since service bureaus can turn presentations into slides in an hour or less. That hour is shrinking, however, because the preferred presentation is becoming the on-screen display on the laptop computer. One advantage of the on-screen display is that it can be prepared and customized very quickly—certainly on a flight, perhaps even at stoplights and in traffic jams for harried sales reps. No longer do they have to use canned, one-size-fits-all presentations—they can now insert personal touches, like used car dealers do when they insert your name into the sales pitch. Such presentations can be foisted on captive seatmates on long airplane flights, balanced between entrée and dessert at lunch, or, preferably, displayed with a high-quality projector on a large screen before receptive viewers.

Just as significant, it's easy to make a boring presentation into a multimedia tour de force. Builds, wipes, and fades are easy to implement, as is adding animated graphics. Even audio and video are practical, though long patches of jerky video may try the patience of viewers. More often, presenters opt for an interactive format rather than a one-way show. Their model is the Internet.

The hypertext format of Web pages is ideal for presentations. Viewers become participants and can follow branches and roots of special interest to them. Though Web formats are constrained by HTML programming, they can include animation and audio/video content and many special effects. When stored and displayed on a local computer instead of accessed over a slow telephone line, images immediately pop onto the screen.

The same presentation can be posted on the Internet, either for internal or external viewing. Modern presentation packages include the option of saving as Web presentations, and readily available conversion utilities can convert existing shows for this purpose.

This format isn't limited to traditional sales presentations, however. The Internet is ideal for all sorts of literature, including brochures, data sheets, technical articles, and reference material. Many companies, especially industrial and technical firms, are turning to the Web format as their primary medium, letting customers access the information and view, print, and store it as they need. This reduces printing costs and inventory, and makes it easy to update material to ensure that prospects always have current information.

For companies concerned about esthetics, the Adobe Acrobat format allows designers to create pages that maintain layouts, fonts, and graphics, but these large files take longer to download from the Internet than conventional HTML-coded pages.

Many companies are planning their documentation as live pages, preparing printed versions from this material as needed. This doesn't always work well, however. One advantage of the Web format is the ability to follow hyperlinks, allowing readers to access as much or as little information as they wish. This concept doesn't work well with printed material, as anyone who ever took a self-directed training course before the age of computers knows. Directions such as "If A, go to page 23; if B, go to 24" are simply unacceptable if you're used to pointing and clicking to get there.

The loss in using hypertext live documents rather than printed material is the consistency and high quality of some traditional printed material. A Web-based document will never be mistaken for a $10 glossy, printed annual report. In truth, many of those documents were poorly designed, rarely used, and eventually discarded. Many companies accept the trade-off in using Web formats to save costs and time and to simplify preparation.

➤ True Value Hardware Brings Home Improvement to the Web

In case you haven't noticed, home improvement is a hot topic today on television, in books, and, yes, on the World Wide Web. Major home improvement stores like Home Depot, Ace Hardware, and Sears all have Web sites, offering homeowners and do-ityourselfers tips on taking care of their biggest asset. The corner hardware store has come a long way in the age of cyberspace. True Value Hardware, with more than 5,000 stores across the United States, has launched a Web site that delivers access to home improvement tips and seasonal to-do lists, and, of course, provides directions to the local True Value store.

True Value Hardware, a subsidiary of Cotter & Company of Chicago, launched its Web site in September 1996. The company has ambitious plans for using the Internet to aggressively drive retail sales, while providing an interactive tool to benefit customers and its retailers.

Giant Step, an interactive marketing and digital communications design firm and subsidiary of advertising agency Leo Burnett Company, designed and maintains True Value's site. Giant Step has built Web sites and interactive content for such popular consumer companies as Walt Disney World, Sony, United Airlines, Hallmark Cards, Maytag, and Oldsmobile.

Giant Step uses a technique called *dynamic construction* on True Value's Web site. In a dynamic site, Web server functions are divided between front-end Web servers that receive hits, and database servers that populate the pages of the site with visitor-specific information.

A visitor to True Value's site might log on with a unique password. The intelligent database server will recognize that last month this individual requested tips on building a deck. It might remind the visitor to stain the deck every year and offer suggestions on top-quality stains available at True Value stores. Or a customer who has indicated a preference for a particular brand of power tools might be notified when there's a special running on that line of tools.

Dynamic construction has two big advantages, according to Giant Step. It gives users a more personalized experience on the site. Dynamic sites are also easier to maintain. The firm changes content once in the central database, and it's automatically changed on every page where that content appears.

To date, the True Value site offers "do-it-your-surfers" access to a variety of helpful and entertaining home improvement categories. One is the FENCEpost bulletin board that connects True Value Web visitors to home improvement questions, experience, and advice. A HOMEwork calendar provides customers with timely tips, seasonal projects, and general home improvement reminders. "This Month At True Value" displays True Value bargains of the month.

True Value's site changes frequently—at least monthly and seasonally—and the available tools make it easy to implement those changes and keep the site fresh. The True Value site is powered by two Compaq ProLiant 5000 servers, each with dual 166-MHz Pentium Pro processors and two mirrored 4.3-gigabyte drives. A third ProLiant 5000 unit contains four 166-MHz Pentium Pro processors and acts as a database server running Microsoft SQL Server and Oracle database software—a task for which the ProLiant 5000's maximized data and input/output (I/O) throughput technologies were designed. All three servers run on Compaq SmartStart's optimized installation of Windows NT Server.

Up until a few months ago, all Giant Step sites were powered by UNIX servers—including True Value's. However, the company converted to Intel microprocessors provided by Compaq, whose ProLiant 5000 performance was comparable to the UNIX machines but for a fraction of the cost. In addition, Web software for Intel-based servers is less expensive. The merchant software for performing commercial transactions over the Internet ranges from $75,000 to $150,000 in the UNIX world, but Microsoft's Merchant Server software is 80 to 90 percent less expensive. It's also easier to install and maintain, which cuts down on support costs. UNIX used to have the most tools available for Web development and production, but now Giant Step says the best tools in the industry today run on Pentium Pro processor-based servers. Microsoft is coming out with everything an Internet service provider wants: mail, news, Internet Relay Chat, FTP, and Gopher, all based on standards. The Microsoft Internet Information Server is an outstanding platform for Web server software, and the Microsoft Internet Studio provides a rich suite for Web site development. Giant Step says that it's much easier to produce and maintain content with these tools, meaning it gets clients' sites to market sooner and can update them more easily.[2]

➤ Selective Promotion on the Web

Like True Value's, most Web pages are designed for broad audiences of prospective customers. While consumer-oriented companies try to use the Internet cost-effectively to promote their products, many firms find it ideal for reaching specific viewers, such as qualified prospects. This ability to address selected audiences is one of the most significant developments since the Internet first became popular. It provides a convenient channel to distribute information to these important audiences.

One example is firms selling highly technical products, including those aimed at engineers. For these engineers, knowing that a product exists and being pitched on its strengths isn't enough. They need to know enormous amounts of detail—often even how to design and test the part. Traditionally, this function has been handled by a combination of trade magazines and newspapers, plus company-supplied printed application notes, product brochures, and data sheets. Now the technical supplier can augment or even bypass that chain, providing complete information on-line at little cost to either itself or its customers.

This doesn't eliminate the trade publication, but it certainly changes its role. For one, it has to analyze and differentiate, not just publish news or technical information an engineer could get more quickly and in more depth on the Web. Just as important, the trade press can filter the flood of data that pours out of the Internet. These roles require a well-qualified editor, suggesting that trade publishers like CMP, with many insightful engineers and other technical professionals who probe and poke, will likely fare better than those publishers who have simply published "scribblings on the backs of ads," which typically are regurgitated press releases.

Obviously, trade publishers are sensitive to this, and most have reacted by setting up Web sites and trying to develop new revenue streams based on the Internet. Ultimately, they will probably attempt to use "push" technology to deliver selective information in imitation of a weekly paper, though probably with a more frequent publication interval.

➤ Reaching—or Bypassing—the Press

Many corporations hope the Internet will free them from the media gatekeepers that have traditionally controlled what their customers saw in print. A good newspaper can help its readers evaluate vendors and their products, but companies don't always appreciate that value. Many would like to bypass that filter.

This is already happening—and not just for trade matters. At one time, the only way most people ever learned business and product news was if it was published or broadcast. Reporters or commentators received their information first from mailed or delivered news releases. In the past few decades, a revolution occurred when paid wire services such as PR Newswire and Business Wire largely assumed that function. They charge companies and public relations agencies a few hundred dollars to deliver electronic versions of the releases to suitable publications and selective wire services such as Dow Jones, Reuters, and the Associated Press. Publications prefer this route since it reduces the mail they receive, is immediate, and provides text in their computers in case they want to use any of it without rekeying. For many years, these wire services, paid and journalistic, carefully restricted their output to the media, maintaining the traditional gatekeeper role of the media.

As the volume of paid releases reached epidemic proportions, however, few made it into print, particularly into the business and national media that most corporations seek. As a result, wire services first allowed individuals and companies to subscribe to their services, then opened them freely to the public. In effect, they've created a new channel that allows companies to talk directly to interested consumers for a small fee.

The ultimate aim of corporations is often to attract direct communication with prospects, of course. Many firms have appealing Web sites and E-mail newsletters and are starting to adopt push technology as well. They want direct contact with their prospects and customers, a modern version of the slick house organ that companies publish for that purpose.

There's a side benefit to the company in talking directly to customers. It provides the firm with marketing intelligence. Companies like National Semiconductor Corporation, which sells to a highly technical audience, can learn a great deal from what engineers request from its Web site. National estimates that 20,000 engineers use its site regularly for data. The firm uses the information it gleans to create leading indicators of overall sales patterns and trends by device type, for example. It combines this information with reports it receives from its sales force automation system to help plan production.

The function of the print publication is not over, however. More new magazines and newspapers appear every day, many slicing the market into narrower or shallower pieces, each trying to find the balance between costly circulation and readers attractive to advertisers. With the flood of information inundating the editors and reporters at these and mainstream publications, it's becoming more and more difficult for public relations people charged with influencing writers to know who's the right contact, much less how to reach them.

Again, technology has come to the rescue—sort of. A number of companies, notably MediaMap, have created programs that largely automate the process of targeting and contacting the press. Using a combination of a dedicated database and contact manager, its programs let users find the reporters at weekly publications who are interested in movies or the editors at engineering monthlies who follow computer-aided design. They also list upcoming special editorial coverage, which publications use to attract advertising but which equally attracts companies offering relevant products and services. The on-line guides give contact information, list special interests, and tell how the editor likes to be contacted; these days, many prefer E-mail, but some still want only phone calls. The programs also help customize and prepare mailings, faxes, and E-mail release of information, create reports, and issue follow-up reminders.

Programs such these have increasingly set a floor of capability and competence in many areas. While not ensuring success, they've become necessities for practitioners. More significant, they handle many tedious clerical and administrative tasks that face any professional, allowing him or her to think more about creative approaches to a job.

➤ Using Technology to Improve Advertising Efficiency

Even advertising is succumbing to automation. For years, companies and agencies have used focus groups to test and optimize ads. It's still an expensive process to produce good ads. Aside from the creative talent required, most advertising utilizes high-quality processes. Virtually all print advertising is generated using sophisticated layout, graphics, and photoediting software such as Quark XPress and Adobe PageMaker, Illustrator, and Photoshop. Video has its own sophisticated technology, some of which is now appearing on desktop computers.

Few companies, however, have tried to automate the production of ads. One area where this has been done is that of the ubiquitous food, drugstore, and other such flyers that are slipped into newspapers every week. Two food chains in Minneapolis, for example, use a program that simplifies this process. Fleming Foods and Fairway Food Division use Z-PIX Datalink ad-planning front-end automatic page-building software to create these ads. Fleming, with 4,000 stores, says it saves $400,000 a year and eliminates two outside suppliers. The sales planning front-end manages store sales plans, item history, and decision support for four ad programs. As sales items are entered into the system, the program looks up their costs, allowances, and coupon information from existing data systems to help managers decide what actions to take. Ad copy, prices, coupon information, and images are automatically inserted into layouts, which are then transmitted electronically to a bureau that completes the layout and produces final images. Using a similar system, Fairway has reduced its advertising staff from 15 to 2.[3]

While the Z-PIX software is designed to supplant an ad agency for routine work, ad agencies themselves also benefit from technology. Ogilvy & Mather (O&M), one of Madison Avenue's most respected advertising agencies, developed a comprehensive database containing market exposure information on thousands of television programs, radio networks, consumer magazines, and other media. The database was created using Informix relational database management system software. Drawing on this database, O&M can maximize its clients' advertising dollars by targeting specific local media for client products, monitoring planned exposures, and comparing the plans with actual exposures.

The advertising business, like many industries, has not been immune from economic restructuring. As companies have downsized and cut budgets, they've also cut their advertising dollars, putting pressure on agencies to justify advertising expenditures. At the same time, the art of marketing has changed dramatically. Advertisers can no longer sell to a mass market as they did in the past. Advertising must be geared to *niche markets*—smaller market segments that demand products tailored to their tastes. To survive and prosper, ad agencies must offer more than top-rate creative talents. They must also offer effective media planning—the art of analyzing and selecting the media that will best reach a company's prospects. Ultimately, agencies must provide their clients with proof that the client's advertising dollars are well spent, reaching their targets, and increasing sales.

In the last ten years, the media departments of many advertising agencies have turned to technology to help them sort through media options, including television, cable TV, radio, newspapers, and magazines. Using spreadsheets, PC databases, and information from data research companies such as Nielsen, MRI, SMRB, J.D. Powers, Arbitron, and LNA, media departments sort through the gigabytes of data to fashion media plans for their clients.

The problem that confronts many agencies is that the proliferation of PCs has left agencies without a centralized repository of data. One planner might create a spreadsheet program for a client, while a second planner creates a second program using different data, and a third planner modifies the first program and uses a third set of data. The result is an MIS manager's nightmare: different program versions floating around, inconsistent data from one program to the next, and no central location for programs and data. Perhaps even more important, the agency as a whole doesn't have one updated central data repository for all planners to access.

O&M decided to establish a specialized group to create centralized applications from which the media planners could build plans. The agency also decided that the new systems would access centralized client databases. These customized programs would be tailored to the special media requirements of the client and, drawing on the central databases, enable planners to create more effective media plans much more quickly than before. O&M chose an Informix relational database that could contain information from many sources, while not limiting the user's ability to query that database in very flexible ways.

One O&M customized planning system supplies both O&M and client planners with a standardized program that provides access to a wealth of information including details such as the size of each audience and its composition for every television program, magazine, and radio network. This information is stored in the database. When planning a promotion aimed at women from ages 18 to 24 in the Los Angeles area, for example, O&M planners can query the database to learn which media will reach the target group, the size of the audience for each television program or magazine, and the cost of advertising. They can then do other queries and compare the trade-offs of cost versus exposure for advertising on a late night TV program, a daytime soap opera, or a niche-oriented magazine. Planners can do in a few hours what would formerly have taken weeks.

When Ford Motor Company first hired O&M, for example, the agency had to handle a TV, radio, and print co-op planning crunch for each of Ford's participating dealers, using the previous agency's method—multiple PC programs and manual inputting of paper-based information. Using this method, O&M planners couldn't keep up with the paperwork. The staff assigned was backlogged, and a doubling of resources was called for to handle the crunch: Seventy media department staffers worked for a week and a half to complete all the plans before the deadline. After the new system was installed, the original staffing level was restored, the process was totally automated, and all media plans are now produced on time, every time.[4]

➤ The Kiosk Arrives

Technology has created another new channel besides the Internet: kiosks. The automated teller has already become a fixture of life, but most still are used primarily for cash withdrawals. Kiosks, by contrast, are used to dispense information. They're becoming popular for shopping malls, discount stores, hotel and building lobbies, department stores—anywhere there's a need for people to get directions and information. Helping to eliminate staff, they can also provide more complete printed, if impersonal, information. The most interesting kiosks provide customer information, perform demonstrations, and even sign up customers and sell products.

Many music stores are returning to the day when a customer could try out records—but in this case, it's with an automated machine like a mini-jukebox that eliminates the possibility for damage or theft of CDs.

A more complicated type of kiosk interactively helps customers find houses and make decisions about complex issues such as mortgages. In Wisconsin, Fox Cities Bank is testing a kiosk called the Home Selector at the Highland Heights Complex in Appleton, while Norwest Mortgage uses the Mortgage Center in the Green Bay and Fox Valley areas. The two related kiosks—both created by interactive multimedia developer Freshly Wired in Neenan, Wisconsin, with agency FH&K Integrated Marketing Communications—are self-directed, interactive products that allow home buyers to learn about housing developments in the local area, view floor plans, and evaluate costs and payment options without a bank representative present.

Like many new applications of technology, the Mortgage Center was inspired by government regulations. The U.S. Department of Housing and Urban Development (HUD) is outlawing payments for mortgage referrals to banks, and the kiosk is designed to encourage real estate firms to refer business in exchange for listing their offerings. Placed in a bank lobby, mall, or other high-traffic location, the Mortgage Center invites potential buyers to find out how much home they can afford and leads them to local listings. It does so without interaction with a loan officer, which many people find intimidating.

The Home Selector kiosk was derived from the Mortgage Center kiosk. It is usually placed in a model home of a development, where it lets buyers learn as much as they want to about the development, including homes that are available, floor plans, and costs. It also provides referrals to the bank. It is interesting to note that the units—without support—cost about half as much as one half-page ad in a large metropolitan daily newspaper.[5]

■ MARKET RESEARCH USING TECHNOLOGY

Another place where technology is substantially reducing drudgery while improving results is in marketing research. The modern corporation uses many types of research, from information on market size and demand through surveys of customers through analyses of sales and distribution. Customer surveys are one of the oldest and most popular means of gaining market knowledge. They're traditionally conducted in one of three ways: a mailed questionnaire, a telephone interview, or an in-person interview. Each has advantages and disadvantages. The first is easiest and cheapest, but usually has poor response, even with an incentive (such as a gift) thrown in. A questionnaire can be designed as a machine-readable form, reducing the time it takes to extract data, but that format doesn't lend itself to the free-form answers that are sometimes most valuable to researchers. A recent variation is receiving and returning the survey by electronic mail.

In-person interviews are costly to perform, particularly if they involve a lot of travel. They work best with skilled interviewers who follow their instincts and experience, rather than simply asking questions and recording the answers.

A common and effective approach to market research is the computer-directed interview conducted on the telephone. The operator, who normally needs only minimal skills, asks questions and records the answers, then is prompted for the next one. Annoying to most people who are interviewed, they may return invalid data. If completed, however, they eliminate the need for separate data entry.

Another approach that has been popular at least partly because of its novelty is the computer interview, for which the interviewee receives a disk containing the appropriate survey questions. Clients of Sawtooth Software, a company that produces tools to facilitate this research, have reported far higher response than with other types of research. The technique eliminates data entry and furnishes quick results and the ability to compile data automatically, making the data usable even before the survey is finished. It also allows complex, branching questions and even demonstrations of products or ads. Sawtooth also offers conjoint analysis tools that allow analysts to evaluate trade-offs between issues.

A newer and even more promising variation of automated interviewing is surveying people on-line through an interactive Web page. This has become so common that many sites routinely collect data on willing respondents, perhaps in exchange for access to otherwise private data. Using a technique known as a *cookie,* companies can also collect what might, in fact, be considered inappropriate data from computer users. Many software and computer product companies even combine an on-line registration form (or a faxed form) with customer surveys.

➤ Mining the Data Warehouse

The most exciting opportunity for research is the data warehouse. Basically a collection of information gathered from inquiries, sales transactions, customer data, market research, outside research, and statistics, the data warehouse utilizes user-friendly graphic analytical tools to let marketing analysts and others who aren't computer professionals dice and slice data to uncover trends and other business intelligence.

The basic data can be stored in a variety of relational databases, many now able to include graphics and audio/video files as well as numbers. Using client/server technology, special-purpose programs collect needed information, which is then downloaded to a client workstation for analysis and manipulation.

A variety of hardware products, database software, and analytical tools are offered by vendors. They range from simple installations on personal computers to multiple-processor systems. Most solutions involve a number of products from different suppliers and generally require significant customization and planning to be useful. Even at that, many haven't met their promise; it turns out that it's a concept easier to appreciate than to implement.

➤ Carl's Jr. Cooked Up a Recipe for Technological Success

These days, it doesn't necessarily take expensive hardware and software to implement a data warehouse. Carl's Jr. did it with PC products. Tracking the number of people who buy french fries with their meals and the percentage of breakfast customers who use the drive-through service can be a daunting task for a company that owns 665 restaurants throughout the western United States, Mexico, and the Pacific Rim. Carl Karcher Enterprises (CKE), parent company of Carl's Jr. restaurants, relies heavily on data from its restaurants to help spot trends in an increasingly competitive industry, in which fast, informed responses to market changes are as critical as fast, friendly service.

CKE depends on Compaq systems to provide its business analysts with timely information where and when they need it. The company first installed Compaq computers in 1994 to support a major shift to client/server computing. Today the company has 20 ProLiant servers distributed throughout California.

Every night, 50,000 to 60,000 records of sales results from the company's business units are collected at corporate headquarters in Anaheim, California. Marketing, financial, and strategic analysts then look at product mix information to gauge, for example, the success of a particular promotional campaign.

The company had used minicomputers previously, but decided to migrate to Compaq servers and Microsoft Windows NT to save systems and operating costs. CKE says that the Compaq ProLiant 5000 provides the performance level of machines that cost two to three times more.

The servers allow CKE to run increasingly sophisticated applications. One of the mission-critical applications it depends on is Essbase, an on-line analytical processing (OLAP) solution from

Arbor Software. CKE uses the Essbase multidimensional database software for comprehensive planning, analysis, and management reporting. Daily sales results are loaded into an Essbase database that combines two years of historical performance information with current data and projected results.

CKE's goal was to improve access to the enormous amount of data it collects, enabling more rapid decision making. Previously, the company needed two days to collect and prepare the data from all its units. Moving the Essbase application to the ProLiant 5000 cut a full day off the process.

Over the next few years, the company's data will grow into the terabyte range. CKE believes that Compaq offers the power, flexibility, reliability, and scalability it needs to reach its business goals. The combination of the Intel Pentium Pro processor and the optimized Compaq host bus, PCI subsystem, and SMART-2 Array Controller gives it the performance it needs. As a result, data collection that used to take 16 hours now takes only 2.

CKE credits the strong alliance between Arbor Software and Compaq for optimizing the performance of Essbase on ProLiant systems. The vendors worked together closely to refine the system's performance. CKE says it just doesn't have time to troubleshoot systems and software.

To help companies like CKE manage an environment of diverse distributed systems, Compaq has introduced next-generation versions of its Insight Manager and SmartStart management and integration tools. CKE says it is confident that the software will make system administration more efficient by providing unparalleled levels of control and consistency.[6]

➤ Tandy Chooses Tandem for a Data Warehouse

One company that has used such a data warehouse successfully is Tandy Corporation. As anyone who has ever gotten a free battery from one of its 7,000 Radio Shack locations knows, Tandy likes to know who you are. The company has a long-standing policy of collecting customer information at the point of sale and, over time, this information had grown into a massive database on Tandy's mainframe computer. It kept five years' worth of records of purchasing activities collected from millions of customer contacts. The problem was how to get at and use this data, since

Tandy's existing system was too slow and difficult to use to help its marketers.

In 1995, Radio Shack senior vice president of marketing, Dave Edmondson, asked CIO Dick Silvers to find a better and faster way for marketers to use the company's most valuable asset, its customer information. It was apparent that the existing mainframe wasn't suitable for the task by itself, so Tandy decided to augment the mainframe with a distributed data warehouse that would make that information more available.

The firm's challenge was to move from expensive mass merchandising to more focused segment marketing. To do that, it created a data warehouse to help it analyze the massive amount of data on its mainframe—2.2 billion rows of information about 55 million customers with 500 million individual contacts and 1.65 billion individual customer attributes. In doing so, Tandy hoped to expand the use of this customer data from historical analyses to marketing and decision support. A step in this process was to shorten the time needed for inquiries.

Tandy chose to offload the mainframe load with Hewlett-Packard HP 9000 enterprise servers, an HP storage system with 2 terabytes' (2,000,000 megabytes) capacity, and a Red Brick Data Warehouse relational database management system with 750 gigabytes of redundant RAID-protected information. This system extracts data in parallel with the existing mainframe. An HP software package, HP Intelligent Warehouse, simplifies the process of querying across the multiple sources of data found in most companies, and Tandy uses Information Advantage Decision Suite for data analysis.

A year later, the system went live. Now end users in marketing and operations research can find answers to their questions in minutes or hours instead of the days formerly required on the mainframe. The timely and detailed marketing intelligence has allowed the company to make the transition from its former mass marketing to a more focused effort, reducing its advertising costs and allowing it to expand its one-to-one relationship marketing with customers.

Having established and proven the system, the company is now planning to expand it to include local store access via an intranet or private World Wide Web site. This will allow local store managers to take advantage of the data as well.[7]

➤ Ameritech Mines Customer Records

Another company with a similar problem—huge amounts of customer data and no way to exploit it—was Ameritech, the regional Bell operating company (RBOC) serving five midwestern American states. With 12 million customers, Ameritech wanted to extract network usage and marketing information from the hundreds of millions of call detail records generated daily, helping to increase its marketing effectiveness and improve utilization of its network. Ameritech selected a 48-processor Tandem's NonStop parallel processing system plus NonStop SQL/MP relational database software as the platform for its customer-usage tracking system. The system collects and collates call records, customer account information, and network usage statistics residing on various legacy systems throughout the Ameritech region. It initially supports 650 gigabytes of data, which company information and marketing analysts will be able to access through easy-to-use Windows-based front-end query tools and report writers. Ameritech's Michael Patrick, director of information systems for its Small Business Services, reports, "This system will help us focus on our customers, reduce cycle time in bringing new products to market, and leverage the sales expertise of our people. It will also allow us to optimize network resources. If we can optimize our decisions by just 5 percent, it will be phenomenal."[8]

➤ New Tools for Analysis

While companies often have to invest significant amounts in the hardware and database software required to implement data warehouses, the data analysis tools used to retrieve information are at least as important in making the effort useful. There are many varieties from many vendors. Even spreadsheets can be used for some of this work, but statistical and special software is required to fully exploit some of these warehouses.

One example is *data mining*, a term defined by the Gartner Group as "discovering meaningful new correlations, patterns and trends by sifting through large amounts of data stored in repositories, using pattern recognition technologies as well as statistical and mathematical techniques."

Statistical software supplier SPSS notes that the process of analyzing data has evolved over the past 30 years. In the 1960s, systems were developed that allowed standardized reporting that provided simple summaries of preformatted information ("What were my total sales over the last three years?"). The 1980s saw the ability to perform ad hoc queries on databases that made it easier to measure performance of specific products or regions ("What were my Midwest sales last January?"). In the early 1990s, vendors developed software that allowed users to "drill down" into their databases on the fly. This allowed users to isolate state sales from a table of regional data, for example. Most recently, companies like SPSS have developed data mining tools that actually sift through databases for meaningful relationships between information. This allows a company to look for clues to why a certain product does better in one place than it does in another.

These data mining tools, says SPSS, lets companies delve deeper into their data, helping them to gain insights and make better decisions. Such tools seek hidden relationships between data, create predictive models, find clusters in databases, review deviations from the norm, and discover associations between activities. This knowledge can help companies to better target their best prospects, identify opportunities to sell additional products and services to existing customers, raise sales performance, and find new opportunities for growth. At the same time, data mining helps retain customers by identifying those most likely to be loyal, while reducing exposure to risk or fraud and allocating resources more efficiently. One company that has successfully used these techniques is Moen, the $600 million manufacturer of bathroom and kitchen plumbing fixtures and supplies that is the leader in branded faucets in America. Recently, the Cleveland, Ohio, company decided to redesign its reporting system to keep its managers and salespeople better informed about sales. The firm sought to consolidate data and improve its integrity, give users better and faster access to information, and analyze and better understand business trends. Finding a way to get reports to users more quickly was crucial.

Previously, a series of monthly and question-specific reports detailing customer ranking, product analysis, and other information was generated by a mainframe-based report writer. Unfortunately, bottlenecks often slowed the reports. Only power users and IT staffers could access information to write a report or run a query. Others had to wait. To correct the situation, the company switched all of its applications to a client/server environment with Cognos Impromptu query and reporting software and its Power-Play on-line analytical processing package. Moen's staff created a data warehouse in Oracle's relational database management system (RDBMS) to store data about sales histories of customers and products, extracting information already in the corporate mainframe environment. Next, the staff built the reporting prototype with staff assistance from Cognos, supplier of the Impromptu query and analysis tools the company chose. The prototype was completed in six weeks.

With the new software, the staff could run through a series of questions and answers without having to run reports; with little training required, they simply clicked the mouse to drill up and down and see sales data in an easy, graphical format. After the demonstration, the project was formally approved and rolled out to users within the targeted six months.

Moen believes productivity has improved because people don't have to rekey information on reports into their spreadsheets to conduct further analysis. With Impromptu, access to detailed data is made easy. Furthermore, since users don't have to be programmers to interface to the data, everyone who needs access has it. All users have to do is know what data fields they want, and they can request them in normal business terms. Salespeople and managers are now able to create custom reports from their own desktops.

The project has been such a success that Moen is expanding PowerPlay's and Impromptu's user base so more people can utilize the tools. The company has also developed a new application to calculate and store contribution margins that will allow users to drill down on any point of the financial model, from gross sales to the bottom line.[9]

■ TECHNOLOGY IN MARKETING

Technology has changed many individual tasks in marketing, but its biggest impact lies ahead. So far, the computer has revolutionized the process of creating presentations and promotional material, but two big changes are only starting to occur. One is the data warehouse—using data already collected by many companies about customers and transactions to perform research and better target niche interests. The second new technology is the Internet, a whole new channel for promotion that allows marketers to directly target special as well as broad interests, eliminating the intermediaries of publishing, retailing, and agents that have controlled their destinies, while creating other, more efficient replacements. When new delivery technologies—which are within reach—provide fast access to the Internet, it's likely to have an immense impact on many traditional businesses, while creating new businesses.

C h a p t e r 5

Putting Willy Loman on Steroids

■ INTRODUCTION

While executives have rushed to add new technologies to improve the productivity of their corporations, they've generally lagged when it comes to sales. Most executives would agree that sales is one of the most important functions of their businesses, yet few of them have focused their own efforts on improving the productivity of their sales forces. It's often because of resistance from their sales representatives, a generally independent lot who are left alone as long as they meet quotas. And in truth, companies often don't target sales for attention because their executives focus on product development, marketing, and manufacturing—more glamorous and interesting operations.

As a result, businesses have been slow to adopt new technology to improve sales operations. Some of the biggest improvements have come in a very piecemeal way, as individual sales representatives have bought—often out of their own pockets—contact management software and even computers on the basis of the recommendations of their peers. For example, about half of the 1.5 million people who use ACT!, a popular contact manager, work in sales.

In many cases, it's taken new businesses with new approaches to really change the way sales are made. Many business trends encourage new technology: more use of the telephone (and eventually the Internet) for dealing with customers instead of face-to-face contact and conventional mail; a move toward direct sales, or at least reducing the levels between a company and its customers; increasing competition that forces companies to reduce costs; and greater segmentation of markets, encouraging corporations to know more and more about their customers. Distribution channels are caught in the middle of these trends. Corporations want to grab more of the eventual selling price, but they're expecting retailers and wholesalers to perform better, too. Retailers are also facing new competition, making them change they way they do business.

One of the most significant contributions of technology to the sales process is blurring its boundaries with marketing and with customer service and support. Sales is traditionally considered a one-to-one operation—one salesperson, one customer. That's true whether it's the $5.15-an-hour clerk at the convenience store selling one package of bubblegum to a neighborhood kid or a superbly qualified consultative sales engineer selling a $10 million super-computer to Boeing.

Marketing, in contrast, is a one-to-many function, with the product marketing specialist typically trying to craft messages that will prepare the way for the salesperson among a wide range of prospects.

Likewise, the customer support person was once considered an aftermarket burden, but more and more, that support or service person has as much long-term impact on sales and revenue as the salesperson responsible for the initial contact.

Technology changes these relationships in many ways. In the first place, the "one" on the selling side of the sales relationship is increasingly a machine, either a corporate Web site or an auto-mated gas pump. It might even be a television set showing an infomercial, with a demonstrator who clinches a sale that is completed only on the telephone.

Second, improved feedback and monitoring systems allow corporations to collect enormous amounts of information on buyers, turning the sales process into a research and even an automated inventory management operation. Technology also holds the potential to help capture more of the sales process, allowing even novice sales reps to benefit from the experience of star salespeople by improving marketing and sales programs.

"In the future, many companies will think of their customer assets as their primary asset," says Microsoft's Bill Gates. "And the world of the Internet has a lot of things where the barriers tend to come down. Distribution systems for lots of products are not as complex, not as much of a barrier. The thing companies have is a customer relationship. It gives them their scale. It gives them their brand identity. And the way they process that information to make custom offers to those customers will be very important."[1]

■ HELPING THE INDIVIDUAL SALES REP

With all the changes in the sales world, however, Willy Loman is still out there pounding the pavement. For high-ticket and high-value sales, there's no replacement for the effective sales representative. Unlike Willy, however, today's outside sales reps are often well-educated, well-trained, and well-equipped for the job. They don't depend on buying martinis and hot lunches for old friends, but on knowledge and a consultative approach. And they use a range of advanced tools to help their performance.

The simplest and the most important tools are probably the cellular phone, the pager, and voice mail. With sales costs per visit approaching $1,000, he—or, increasingly, she—can't waste time and travel. These basic communications tools let reps respond immediately and change plans quickly if needed.

Next in value is contact management software, today's high-tech Rolodex or Day-Timer. These programs have more functions than a Swiss Army knife and are far more useful to the salesperson. Starting with a comprehensive telephone and address feature that can track spouse's names, birthdays, and even a favorite ball team, these programs include scheduling, automatic phone dialing, the ability to generate follow up letters quickly, and reminders of things to do. They run on desktops, laptops, and palmtops, with automatic synchronization for people who use more than one device. They can even tie into corporate systems to avoid scheduling conflicts and to update records.

A contact manager is useful to any salesperson, but more and more outside sales reps use portable computers for managing and maintaining contact with their customers and employers or principals. Many carry powerful but bulky laptop computers. Others choose smaller but limited handheld organizers such as the U.S. Robotics Pilot or tiny computers from companies like Hewlett-Packard, some running a minimal but compatible version of Windows called Windows CE. Many of these small computers include E-mail, and, as wireless E-mail becomes widespread, their usability will grow significantly.

These small computers are generally used for limited purposes; composing lengthy messages is impractical, for example, and even if they include spreadsheet programs, not many users are likely to spend much time using them in this manner. These simple products, however, have succeeded, while the much touted earlier personal digital assistants (PDAs) such as the Apple Newton found limited markets. The handwriting recognition that attracted so much attention to the Newton is of limited use for writing letters—especially if it's picky about the way you write—but it, or the miniature keyboards some include, is fine for scheduling appointments.

➤ Contact Management Helps Argo Manage Sales

Argo International Corporation, a worldwide electrical and mechanical distributor, inventories electrical and mechanical equipment and parts, including electric motors, switch gear, drives, pumps, compressors, and a wide variety of peripherals. Headquartered in New York City, it maintains one of the world's most complete inventories of mechanical and electrical equipment from various manufacturers in a growing network of 20 fully stocked warehouses located throughout North America, Europe, and Asia. It also has 15 U.S. branch locations and 10 international branches. The company has been in business since 1952 and currently has approximately 250 employees.

Although most people consider contact management software to be a personal program, Art Johns, the sales and marketing manager at Argo International, knew that contact management should be done at a company level. Like many companies, Argo had a computer system used primarily for accounting. It wasn't geared toward recording pertinent data about customers. The company was not able to leverage its customer information to support the sales force and drive its business.

Argo decided to do a better job at utilizing available technology to improve the sales effort. Its salespeople didn't have an efficient way of tracking electronically what was happening with its customers. Because of this, the firm wasn't able to data mine, organize mailings, or telemarket effectively. The data was locked away in each salesperson's black book.

Argo needed the capabilities of contact management software to operate more efficiently, but it needed backing from the management team and the sales force to change the company's procedures. Johns convinced management that Argo's customer information was its most valuable information and that the company could use it creatively to drive sales and, ultimately, profits.

He was familiar with ACT! from a previous position and believed it was the right product for Argo, but he also realized that he had to roll out the software in stages to build support. Several sales reps were chosen to test whether it would actually help them and to help determine whether Argo would make a larger commitment. The sales representatives who served as guinea pigs said that ACT! helped them with time management and follow-up, and in communications to customers and prospects through mailers and phone contact.

After the successful test, Argo began rolling out the product to three sales offices with suitable personnel, computer experience, and willingness. It decided to wait for ACT! 3.0 before it went ahead with wider use, so it would have the latest version with features such as linking.

Johns also spent time developing a customized format for the contacts so that the information would be standardized throughout the company. He wanted to make sure that when the rest of the company used ACT!, the information could be easily combined into a master database to be used in marketing. Argo used the Castle Group to help in this customization and in training.

That training was another key to success. Once salespeople were trained and entered their contacts, most found ACT! very useful. They've found it helpful in managing itineraries, uncovering new accounts in old areas, and recording activities and follow-ups. Others like its time management. Instead of carrying around a pile of notes reminding them of things to do, the reminders are all in one place.[2]

➤ Mining Contacts with GoldMine

Data Base Inc. (DBI) is a leader in computer data protection with facilities all over the United States. Since 1976, it has helped protect computer information for thousands of companies and today protects information assets of nearly 3,000 companies and organizations. Its customers include nearly half of the Fortune 500 companies, five of the six top commercial banks, and four of the five top telecommunications organizations. DBI's 30 sales representatives work from eight offices around the country. They use GoldMine contact management software to keep track of their customers.

The sales force previously lacked an integrated method for sharing information and coordinating activities between sales reps and management, causing them to frequently duplicate or overlook calls to prospects. In early 1994, DBI decided to overhaul its sales process to enhance communication within the sales and marketing departments. It also wanted to get the company's executives and management more involved in the process.

DBI evaluated several packages but found that GoldMine best fit its style of business with its sales automation features and synchronization capability. Now DBI's sales representatives synchronize their data daily with regional offices, which synchronize weekly with headquarters. This regular synchronization provides the company with a good picture of contacts and sales. The company's vice president of sales spends about 70 percent of his time on the road. He used to spend hours each week calling every representative for an update, but the frequent updating of data with the software lets him spend more time assisting his sales representatives.

DBI began seeing benefits immediately after adopting GoldMine. The company estimates that it saves the sales staff 60 hours each week—an additional 3,000 hours spent selling each year. Before the company installed GoldMine, each sales representative had to write three reports per week, a process averaging two to three hours per week each. Now it takes just minutes. Even small benefits, such as eliminating redundant typing when exporting contact information to Excel spreadsheets or Word documents, have increased productivity.

DBI is using more of the program's features as it becomes familiar with the software's capabilities; in the past, when a rep met with a prospective client to prepare a quote, the process was tedious and slow. DBI solved this problem by developing a pricing module and linking it to GoldMine. This module uses a bidirectional link, which transfers data into Excel for calculations, then sends it back. A sales rep can now go to a client's site, input 10 to 20 variables, then use Excel data to generate sales quotes and reports in GoldMine.[3]

➤ Sales Force Automation Beats Sales Rep Automation

Though ACT! and GoldMine have some group capabilities, they are really designed primarily for individual sales reps. Many corporations are adopting more comprehensive sales force automation that manages more of the sales process. A big problem with enterprise sales force automation, however, is getting sales reps to use it. MIS departments hate individual contact managers, which salespeople like, because they can't capture or extract information from them, but salespeople often won't use the more structured overall systems imposed from above. Forrester Research estimates that, in fact, 50 to 80 percent of high-end sales force automation schemes fail. Besides cantankerous but successful salespeople, who may simply leave the company if forced to use a difficult scheme, many such systems are too ambitious and take too long to implement. Perrier, for one, chose a medium-level automation package, SalesLogix, that's basically a shared corporate version of a contact management package with added features important to sales.

SalesLogix was developed by the man who created and built ACT! to a prominent position in the market. It allows people and companies familiar with ACT! to use a familiar format and existing data, yet implement a system designed to serve the need of moderate-sized companies instead of individual salespeople. Other companies use other products; some even develop their own.

At Holiday Inn Worldwide, approximately 200 sales representatives stand between hotels and travel coordinators, agents and corporate travel managers. The hotels themselves are franchises, so the sales force doesn't commit to prices or dates, but manages contracts and contacts. Obviously, there's the potential for a lot of overlap; most organizers and travel coordinators need to deal with many properties, and many organizations have multiple staff members working on booking rooms.

Because of the structure, the sales reps can't compete with the traditional sales tool—lower prices—but have to offer improved service instead. In the past, the reps had masses of information buried in cabinets, including contracts that could expire at any time. Not being able to locate that information quickly hardly impressed clients, and sometimes it created big misunderstandings. Yet some reps manage as many as 700 contracts with airlines for more than 3 million room-nights a year.

If it was difficult for the salespeople, it was even worse for their managers. They couldn't tell what was going on, even though the reps spent valuable time filling out reports. Definitive data on trends and comparisons to competitors was difficult or impossible to obtain. Managers didn't know if they needed to hire more sales reps or redirect the efforts of those on staff. To improve the situation, Holiday Inn worked with Computer Sciences Corporation to develop a better system.

Now all account managers carry notebook computers with custom interfaces that let them access client contracts, competitive information, airline codes and cities, and account information. It also includes individual scheduling software so reps can manage their time effectively. These notebooks can be connected remotely or in the office to an information system to coordinate information, allowing the whole organization to work together effectively. The system reduces paperwork, the bane of most sales reps' lives. And if a sales representative leaves, his or her data is still available to the organization rather than walking out the door with the rep. Not that that many leave—the system has improved productivity and morale, reducing turnover. The system also lets managers better monitor the sales process, allowing them to make better decisions.

As do most sales force automation products, this application requires the reps to carry laptop computers. The reps can also use them as multimedia catalogs and to give compelling demonstrations and presentations, for composing letters and E-mail, and, in many cases, for actually filling out applications and ordering products.[4]

Among the areas in which this is proving to be a time-saver is when applying for life insurance and real estate mortgages, as well as for ordering complex products like large trucks or computers, which offer many options, some incompatible.

■ ENTERPRISE SALES FORCE AUTOMATION

A whole new class of enterprise software companies supply so-called front-office software that helps manage the sales process. In a much larger sense than ACT! or other individual contact managers, products like ClearSales from Clarify organize the sales cycle, from managing leads, contacts, and activities to producing quotations, forecasting, managing accounts, and even handling ongoing customer service. From wherever they are, sales reps can communicate with the factory (and customers), sharing valuable information.

One front-office software firm that has been particularly successful is Trilogy of Austin, Texas. Chrysler uses its Selling Chain software to help dealers share product information with customers, allowing them to pick exactly the car and accessories they want with links back to Chrysler to provide feedback and manufacturing information. IBM uses the product to help 35,000 sales representatives who use laptop computers to configure complex personal computer products and to automatically enter orders, initiate billing, and report on the transactions. The system fully integrates what were once individual unconnected parts of a company's business, joining them to the firm's existing financial and manufacturing processes. Other front-office software includes opportunity management, help desk, catalog, configuration, pricing, quoting, and order management software.

The key to the success of these products is that they don't simply automate discrete, existing tasks, but fundamentally change the way a company approaches the sales process. As such, they have to be built into a new operation or become part of a company's overall reengineering effort. And though many people advise corporations to become more customer-oriented, as these products demand, it's still a difficult move for many companies.[5]

➤ Selling Car Rentals Using the Internet

The newest aid to helping salespeople is the Internet. Alamo Rent-A-Car is the nation's largest leisure auto rental firm and it was the first in the industry to allow customers to reserve cars directly over the Internet. It also found another use for the Internet, helping its own sales force sell its services to travel agencies and corporations.

Alamo had a wealth of critical inventory and pricing information, but it was unavailable to its sales force who needed it in the field. The company needed a comprehensive collaboration system that would allow its national sales force to access this critical corporate information so that they could customize contracts and special packages for travel agents and corporate travel planners.

Alamo chose IBM to develop a sales force automation system incorporating Lotus Notes to form a fully networked mobile sales force. The car rental firm created an intranet for the sales force, then equipped it with IBM ThinkPad portable computers to allow the sales representatives access to a CICS mainframe database for critical account, inventory, and pricing information. Using this intranet, Alamo's sales team can access information while visiting corporate sites or travel agencies to create custom packages based on current price and inventory information. Alamo also intends to equip its contract sales employees with access to the CICS database through an extranet using Lotus Domino, giving this part of the sales team access to the same critical information.

The rental firm has reaped solid benefits from implementing a sales force automation system, resulting in faster and easier turnaround of business-critical information to the sales force and allowing the sales team to better pursue prospective clients and better service current contracts.[6]

■ CALL CENTERS HOLD A BIG STAKE IN SALES PRODUCTIVITY

Most large companies use individual outside sale representatives, but there's definitely a trend toward less costly forms of sales. One of the biggest booms is in call centers that turn telephone calls from customers and prospects into orders. Figure 5-1 indicates a doubling from 1996 to 2001. These operations have many efficiencies over face-to-face contact, notably the shorter time to enact a transaction, the ability to connect a customer anywhere with a product specialist at a different location, the ability to locate in low-cost areas of the country, the possibility of hiring a typically less qualified (and less well paid) sales force, the ability to deliver proven scripts to customers, and easy access to product information.

The technology for these call centers has become very sophisticated. These products go far beyond distributing incoming calls to agents. They utilize advanced computer-telephone integration that allows the agent to sell and enter orders efficiently, while providing management information that allows it to react quickly to changes in demand.

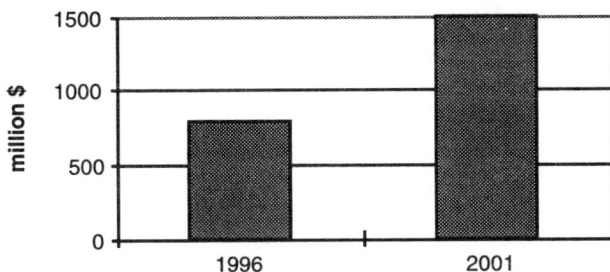

Figure 5-1. The U.S. call center market is expected to double between 1996 and 2001, according to Dataquest.

These centers operate in two fundamentally different ways. One processes calls people make to 800 numbers. The other is the dreaded outgoing sales center that always seems to call when you're eating dinner. The same equipment can be used for both, but the person who makes an unsolicited call to try to sell mortgage insurance for a new bank loan is not likely to be the one who receives a call from someone applying for the loan.

➤ Call Centers at a Bank

Banks, in fact, are perfect examples of subtle call center selling. People who call banks want something, but it isn't necessarily something that will make money for the bank.

It's a classic case of the thin line between customer support and sales: If someone wants an account balance, the bank can try to get the person to use an automated system, or it can take the opportunity to pitch a certificate of deposit or sign the person up for a Visa card. The bank certainly doesn't want the customer to simply take up an employee's time on the phone or, even worse, in person at an expensive branch office. Banks are faced with changing the way customers act—and in the process changing the attitudes of their own employees.

One bank that has successfully faced the challenge of using technology to help implement a change in culture as well as function is BankBoston, a holding company that includes the First National Bank of Boston, Bank of Boston Connecticut, Rhode Island Hospital Trust National Bank, and BayBank. The fifteenth-largest bank holding company in the United States, it has assets of $62 billion, more than 500 branches, and two call centers.

Besides the issues involved in merging the operations of these different banks, the bank had an outdated system that required its tellers to wade through layers of menus to accomplish even simple tasks. The call center used a completely different system, and its agents couldn't access information that tellers could and vice versa. The bank was strong in corporate and international banking, but it wanted to strengthen its consumer business and felt it had to improve its responsiveness and service culture. The old teller and call center systems had to change.

A key part of this effort was to develop a new graphics-oriented retail workstation used by all associates, including those at the call center, to access customer and account information, and a separate teller workstation for handling money. It's a classic client/server application, with considerable power and processing done locally on a personal computer, unlike the host-based text system previously used. The application server runs Argo Bankpro software, which interfaces to the bank's mainframe computers. All the nodes are linked together in a branch over a local area network and overall via 56K digital circuits and routers.

Within a three-year period, the bank successfully developed and installed the new system, including the unanticipated addition of 235 BayBank branches.

The systems used by associates now access all required information, allowing them to enact transactions, such as new account and loan applications, quickly. Laser printers at each station allow associates to stay with the clients, and preprinted forms were replaced by templates stored in the computer. Even standardized letters and confirmations were created to save time and provide a consistent response.

The loan application process was so streamlined that associates could process and print the form in a few minutes; previously, it had to be printed at a central location and mailed, a process that took days. Even customers making deposits have shorter waits in lines since each transaction is quicker and, as a bonus, customers get clear, printed receipts, not hard-to-read, handwritten carbon copies.

This project was the largest technology project ever undertaken by the bank, both in terms of budget and staff time, and the most aggressive in schedule. It is interesting to note that the process was triggered when the call centers and branches were combined into the same organization, ending the second-class status the centers had previously endured.

This step reflected the bank's early (in 1993) sense of the changes ahead in the banking business. Even integrating the branches into the overall corporate wide area network changed the culture, allowing the bank to become more aggressive in offering and delivering new products to its customers. With the new system, it's easier for everyone who has contact with customers—and that's everyone in the bank—to be more responsive and sales-oriented. Selling no longer means you'll have a lot of trouble and paperwork; instead, it becomes easy.[7]

➤ Telephone Call Center

A similar change from *support* to *sales* was driven by technology at Bell Atlantic. There, telephone agents who once simply took requests and answered inquiries are actively selling services while performing traditional tasks more efficiently.

The giant telephone provider created what it calls the SaleService Negotiation System to handle requests for telephone service, change of service, and disconnection of service. But it also handles customer accounts, inquiries, and maintenance while integrating customer address validation, service connection, and number assignment.

Most of all, it does this through a process of negotiation with the customer. A script steps the agent through a dialogue that discloses all regulatory information—a complex headache for the telephone company as well as for the customer these days—and prompts the agent to offer appropriate additional revenue-generating services such as call waiting and voice mail. In the process, it avoids conflicts, such as services that aren't available in certain locations because of equipment limitations. On-line help lets the agent answer even obscure questions. Bell Atlantic claims the system helps it address three challenges it faces: customer marketing, legacy system functionality, and the sales process.

These days, there are a wealth of products available to consumers and small businesses from the local phone company, but it's often difficult to present them effectively. Bell Atlantic's system tries to do that, even taking clues from a customer's requests to suggest additional services.

The phone company has a problem few businesses face. Some of its products still maintain century-old compatibility and many legacy systems are difficult to use and understand. In the past, for example, the agent had to call another phone company employee to reserve a new phone number, putting the customer on hold while the other agent queried *his or her* computer. Now the agent can do that on-line. Likewise, the agent can check credit, verify service availability, even schedule installation during the same call. Likewise, the options that had to be selected by the agent were so complex that the process was long and annoying to the customer and often resulted in errors and lost sales opportunities. Now it's just a point-and-click process.

The new system allows agents to tell customers about additional services that meet their needs, raising revenue. It also reduces customer service representative training and frustration, as well as errors, contact time, and credit problems. Furthermore, it improves customer satisfaction. Perhaps one day jokes about calling the telephone company will go the way of black phones with rotary dials.[8]

Of course, there are many other companies that use call centers to sell products. Many take advantage of another technology—television—in an innovative way. Infomercials have become such an effective marketing tool that many television and cable networks broadcast nothing else. And all these infomercials either sell products directly via 800 numbers or ask viewers to take some other action, such as requesting information, again over the telephone.

➤ Call Center Management

One of business's largest call center operations is that of MCI's Consumer, Sales, and Small Business Division. To manage its 14 centers across the country, MCI uses TotalView workforce management system software from IEX. TotalView, a Windows-based client/server system, is connected to an MCI network of Rockwell Galaxy and Spectrum automatic call distributors (ACDs), which route incoming customer service and sales calls to more than 3,000 agents nationwide. With the capability to record the status of each individual call center, as well as maintain an overview of all 14 locations, the program provides MCI with a comprehensive picture of its call center operations. MCI says that TotalView lets it view and treat its 14 locations as one. By establishing schedules at a national level and allowing each site to individually monitor the status of calls and agents, the software handles MCI's 5 million calls efficiently and provides more stable and consistent schedules for MCI representatives. It provides MCI with extensive scheduling and forecasting tools.

TotalView lets businesses, including financial institutions, public utilities, hospitality companies, reservation centers, and telecommunications carriers, optimize their call centers while cutting costs. In one system, it integrates forecasting and scheduling, attendance and agent productivity tracking, payroll input, vacation and holiday planning, meeting scheduling, and adherence reporting.[9]

■ INTERNET SHOPPING IS BOOMING

While call centers have already revolutionized the sales process for many companies, observers see the Internet revolutionizing shopping once again. Still in its infancy, the Internet already has created some sales success stories and more are sure to follow. Once a few important issues are resolved, it's likely to become a major sales channel.

Forrester Research expects the volume of Internet commerce to explode, as shown in Figure 5-2. The Internet offers many benefits to retailers: no expensive real estate, structures, or displays; minimal sales force; the ability to deal equally with customers in any geographic area; an interactive multimedia catalog that can provide as much information as a customer could want; no expensive catalogs and postage; instant communications; instant updates for price changes and stock level; and the ability to customize and offer specials to customers based on their interests. Basically, a firm can offer products for sale at minimal cost. Many don't even maintain inventory, but simply drop-ship packages, often for overnight service.

On the other hand, there are still some problems. One is that not that many people are actively using the Internet—far more

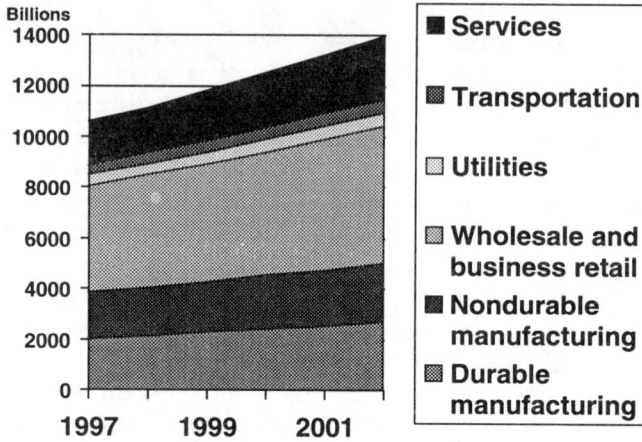

Figure 5-2. U.S. Internet commerce should grow dramatically. *(Source: Forrester Research, Inc.)*

have phones and mailboxes! And of those that do use the Internet, few have high-speed connections, challenging merchants to entice viewers to wait while snappy graphics—even video and audio—load. Many people aren't comfortable providing personal information over what is widely known to be an insecure system, though that should change as security and encryption schemes improve and customers gain experience.

Many of the companies that have pioneered sales over the Internet have been new firms, but some existing merchants have been successful, too. At this point, it's definitely a specialty game and most retailers still consider it as much of a test, even a promotional channel, as a business. Nevertheless, a recent study by Computer Sciences Corporation and *Retail Info Systems News* found that 20 percent of retailers in 1997 offered on-line sales, up from 11 percent the year before, and almost 40 percent expect to offer sales on-line by 1999.[10] Cowles/Simba Information expects consumer on-line revenues to grow dramatically, as indicated in Figure 5-3.

Some of these sales efforts are less sales than promotion, however. Volvo, for example, lets consumers "design" and price their ideal car and accessories over the Internet using Trilogy's Selling Chain for the Web software, but there's no space for them to enter their Visa card number on the page; they still have to go to their local dealer to buy a car.[11]

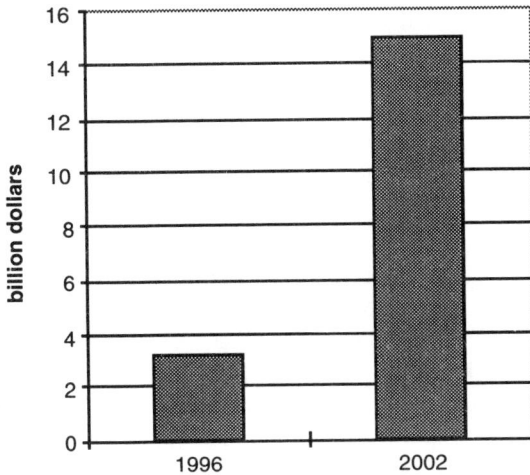

Figure 5-3. Consumer on-line market revenues will increase from $2.5 billion to $15 billion, according to Cowles/Simba Information.

➤ Selling Consumer Products on the Web

Two companies that have succeeded on-line found niches ideally served by the Internet. One is Amazon.com, which provides an extensive on-line bookstore. This business seems especially well suited for the Internet, for while many people enjoy browsing in bookstores, even the largest bookstores found in large markets can't stock a fraction of the books that are published. And some of the people who most depend on books live in places far from big bookstores.

Amazon.com is more than a listing of 2.5 million books, however. It offers reviews and links that encourage its customers to linger. For special-interest books and limited-audience authors, Internet sales can be a godsend, for it is very difficult for them to find shelf space among all those best-sellers and cookbooks.

Amazon.com doesn't maintain inventory, but it has publishers or distributors drop-ship the books to buyers. It offers significant discounts, though shipping and handling charges can eat up that advantage. Amazon.com's revenue has skyrocketed to $100 million after only three years, leading it to a highly successful initial public offering (IPO). It's also attracted large booksellers like

Barnes & Noble and Borders Books & Music, which have a challenge Amazon.com doesn't have: how to sell books on-line without cannibalizing their heavy investment in bricks, mortar, and inventory.

Borders also opened its virtual doors on the Internet, putting its vast book and music catalog in the hands of electronic shoppers. Borders is making the move to the Web to improve customer access and demand for special-order book and music titles. Unlike most other virtual book and music sellers, Borders is a national-scale retailer, with an extensive inventory and distribution infrastructure already in place. Since opening its first bookstore in Ann Arbor, Michigan, 26 years ago, Borders has seen its sales soar to over $1 billion per year, while its locations have grown to 200. Borders Group is the world's second-largest retailer of books, music, and other informational, educational, and entertainment products.

Internet and intranet access will let Borders satisfy special-order requests, both in-store and on-line. Borders selected IBM Net .Commerce to get it on the Web. It's a service that makes it easy for merchants to establish on-line storefronts with the industry-leading Secure Electronic Transaction (SET) security protocol, intelligent cataloging tools, and extensive database access.

"The Internet is bringing a whole spectrum of entertainment options into the mainstream with greater speed and ease than ever before in history," says Philip Semprevivo, CIO of Borders Group. "Borders plans to take advantage of these electronic delivery systems by making it as convenient and interesting as possible for customers to shop for and buy books and music on the Web."[12]

➤ Ordering Food on the Web

Another new type of business is an ironic throwback to an earlier era when people ordered groceries over the phone—or even by mail!—and the stores delivered them. In those days, the orders were for ten pounds of flour or a dozen eggs. Now supermarkets stock tens of thousands of items, and customers must choose between many similar items from different suppliers, in different sizes, and with different nutritional preferences. The only way to choose is to walk through the store, where you're likely to be distracted by impulse items or to maintain a list of preferences.

The consumer also has a different motivation than previously, too. Many of today's busy workers have more money than time, and they choose to use what extra time they have for pleasure, not grocery shopping.

The two trends seem ideal for Internet grocery shopping, and that's just what Peapod offers. In conjunction with major grocery retailers in metropolitan areas, it allows shoppers to choose and order on-line for convenient delivery. The consumer can order while watching *Seinfeld* and have the groceries arrive after work the next day. There is a monthly fee (about $6), a delivery charge (typically $5), and a commission of 5 percent of the order. The average order is about $110, so the cost is about $10. That is partly counterbalanced by avoiding impulse items and by taking advantage of discounts Peapod offers.

The service is an attractive one for food suppliers, since it identifies targeted buyers for their products. One of the features of the service, in fact, is that it collects large amounts of data on its customers, which it uses for promotions and to suggest future purchases.

The supermarket chain that partners with Peapod has little involvement in the process, since the on-line service takes the orders, collects the products, and delivers them to the customer without a need for additional store personnel or equipment. Ultimately, it could result in significant sales.

Peapod started with one 486 computer and 400 customers, grew to separate Hewlett-Packard HP 9000 minicomputer servers in the second market it entered, then consolidated its computer operations to take advantage of the economies of scale. It maintains two large Informix databases—one for its orders, one a data warehouse that stores customer data. The service now has about 40,000 customers and has been growing rapidly.[13]

➤ A Sharper Image on the Web

Another innovative business, though one well established in storefront and catalog sales, is Sharper Image. The merchandiser of upscale toys and gadgets has enjoyed great success selling new and unusual products, and the Internet seemed like a natural for its target market. An early presence on the Web, Sharper Image, like most firms, began with an on-line version of its print catalog.

After a meeting between its founder, Richard Thalheimer, and Steve Jobs, founder and chairman of Apple Computer, which acquired NeXT Computers, Sharper Image changed its philosophy completely. It now generates Web pages on the fly from a NeXT NetObjects object-oriented database of its merchandise. This way, there is no need to laboriously create HTML-coded static pages, then update them as changes occur in pricing and features. The company only needs to maintain its product database. It also fosters interactivity, allowing customers, in effect, to create catalogs of items they're interested in.

Amazingly, the system was implemented in only 60 days, and Sharper Image claims it paid for itself in 90 days of operation. Overall, the company's on-line sales in 1996 were $1 million, four times those the year before. It attracted ten times as many visitors in 1996 as in 1995, an indication of the growing interest in shopping on the Web.[14]

➤ Buying Computers and Software on the Web

Sales and distribution of software are being rapidly transformed by the Net, too. Both consumers and businesses can download software immediately instead of waiting for new products to get to their local store or even through the mails. With today's technology, companies can post promotional demos that expire after a certain time or usage, or they can sell full versions using credit cards and passwords that unlock them. The products come with only on-line or electronic documentation, but so do many retail products, and the savings in production, inventory, and shipping are significant. The Internet also allows instant upgrading or bug fixes—though with today's huge programs, it can sometimes take hours at the access speeds most consumers use.

One effect of the growth of electronic distribution of software is a change from the idea of buying a product toward that of a subscription. A new version arrives every six months, providing new features (and corrections of errors) for the user and a virtual annuity for the supplier. With this change, pricing patterns have also been changing, from $500 for one-time purchase for popular business programs to around $100 per year for upgraded versions.

At the same time, electronic software distribution is broadening the market, because smaller firms can deliver products easily at low distribution cost. Some, like Qualcomm, offer free basic versions of products such as its popular Eudora E-mail reader, hoping to hook customers who pay for upgrades. Netscape allows individuals to get copies of Netscape for free, asking them to pay, but really depending on the technique to build its large market share, as it tries to sell browsers and business products to organizations.

Microsoft, of course, simply gives away Internet Explorer, trying to increase its market share. Other companies offer shareware, depending on time limits or guilt to get customers to pay. Some of these programs are small, special-purpose or special-interest products, whereas others are capable products equal to those from conventional suppliers.

Even for big companies that can garner retail space and wide customer attention, electronic distribution of software is appealing. Microsoft has signed a deal to deliver its products to government users via software.net, an Internet-based software superstore that focuses on software distribution. software.net offers about 22,000 software products altogether, with 2,200 available for immediate downloading. The government deal covers $50 million in software for the Defense Logistics Agency and the Department of Defense Procurement Agency, but saves the government $30 million in packaging, shipping, installation, and maintenance costs. Users will access the software from password-protected sites. The deal also covers support from a Microsoft consulting group.[15]

➤ Travel on the Web

A highly visible Internet success is another of those merged sales-support functions that's typical of the Web: the consumer version of American Airlines' wildly successful and widely emulated SABRE reservation system. A natural evolution of the original system developed for the airline's own use and for travel agents, it's a prototype of an approach that could eventually have a massive impact on the travel business.

Airlines don't pay travel agents high commissions—typically 10 percent or less—and in the past, it was worth the small amount to shed some of their huge costs for processing individual reservations and tickets. With automation, first in their own telesales centers, and even more so in Internet-based systems that require no human intervention, it's becoming more and more attractive to sell tickets directly to consumers. They're already reducing commissions on small sales, putting severe pressure on small travel agencies.

As a side benefit to the airlines, airline computer systems aren't likely to suggest alternative lower-cost approaches or other carriers, as travel agents can, ensuring that the airlines capture greater market share and profits. By tying the systems into frequent-traveler systems, they provide an efficient sales and marketing system that engenders customer loyalty and commitment. Consumers may benefit, too, since they can view options if they expend a little effort, finding more convenient times and better fares. Forrester Research expects on-line ticket sales to grow to $10 billion by 2001, as shown in Figure 5-4.

➤ Business-to-Business Sales on the Internet

Consumer sales on the Web get the attention, but the Internet is probably having a bigger impact on business-to-business sales. Patterned on the airline reservation system, pcOrder was formed to provide a giant marketplace where computer dealers and large corporations could shop and order personal computers and parts efficiently over the Net. It lists more than 150,000 products from over 800 manufacturers from a database that is updated constantly. In addition to the actual sales function, this system provides current pricing and product information, a boon in a business that changes so rapidly. The system is already receiving more than 80,000 hits per day, while providing millions of dollars in quotes.

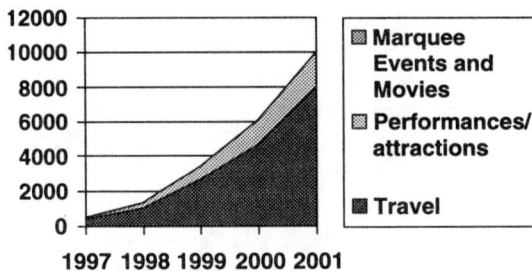

Figure 5-4. On-line ticket sales in the United States. *(Source: Forrester Research, Inc.)*

➤ Entrepreneurs Bring the Car Lot to the Desktop

Buying a car is an unpleasant, anxious experience for many people, but the Internet has the potential for changing the way car dealers and car buyers interact. An on-line car shopping service, Auto Town, strives to make car shopping a pleasant experience.

Auto Town blames a lot of the fear and hostility car buyers feel to a lack of control and negotiating expertise: A car buyer typically has less negotiating experience than the car salesperson who bargains all day long. Auto Town claims to put the customer in control by providing a place to browse through new and used cars from the convenience of their homes, so when they walk into a dealership, they are not walking in blind.

For example, if a customer wants to buy a used car, he or she can do a search on Auto Town and see what is on a dealer's lot. The customer can then walk in with some confidence and say, "I saw a '92 Buick Century with 22,000 miles on. This is the stock number, and I would like to see it."

The power behind Auto Town is a sophisticated Oracle-based search engine that allows car shoppers to define their own searches. A search can be inclusive, or it can be limited by region, budget, make and model, or mileage or age of the vehicle.

Auto Town wanted to provide a service to meet the needs of car dealers as well as customers. It wanted to bridge the gap between customers and dealers. The company marries the information customers need with the tracking and return-on-investment (ROI) reporting that dealerships require. The value of the database to the customer is the search engine. The value to the dealer is customers—and proof of their advertising investment. Car dealerships have traditionally invested heavily in traditional forms of advertising—newspapers, radio, and television ads—without knowing how many potential customers they actually reached. There was no way to track the ROI on those advertising dollars. Auto Town provides a form of advertising that will not only track response but will tell how effective the investment was. Auto Town relies on "Web mining" to develop the ROI reports for advertising dealers, sifting and analyzing large amounts of data.

Auto Town also acts as a value-added service for dealers, giving them additional capabilities to extend to car buyers. For example, people can enter parts orders or service appointments via the Web site.

To create the service, Auto Town sought a product that would grow with the company and mesh well with the Internet, a very distributed environment that is changing continuously. It chose Oracle technology to provide comprehensive searching capability, a large capacity for data storage, and rapid response. The company believes that its Oracle database server is capable of handling that challenge.

Auto Town runs the server on a Silicon Graphics (SGI) Challenger server with a 9-gigabyte database that is growing daily. Its Webmaster uses an SGI workstation to develop and manage the database and the site. Dealers who advertise on the site use their own computers—Windows-based PCs or Macintoshes—to access the Auto Town system for frequent updates. The firm has provided dealers with an infrastructure that allows them to post changes to the database without having to wait for someone at Auto Town to update it. Dealers have to update their inventories when a car is sold, but this also means that their new inventory is available quickly to Web shoppers.

Individuals can also post their used cars for sale with Auto Town. The company notes that the used-car market is twice the size of the new-car market, with 70 percent of the market being individuals selling to other individuals. Auto Town says that most of the competition is focusing on new cars and new-car sales, which is not where consumers spend their money.

Auto Town plans to expand to all 50 states in the United States and internationally to Canada and Europe. With an eye toward electronic commerce, heavier traffic, and expanded information on the site, Auto Town's executives expect a wild ride. Accommodating dramatic growth shouldn't be a problem, since it's a virtual company that doesn't face many of the limitations of traditional companies. Because everything's electronic and its procedures are automated, Auto Town claims it can grow much faster than other forms of company.[16]

➤ Communicating with Dealers

Getting current information to stores is a major problem for many businesses, and the Internet provides the perfect channel for communications. Adopting security to keep this information private has led to hundreds of *infranets*—private virtual networks connecting welcome users while excluding others. General Motors, for example, has 8,500 dealers, and they need a huge amount of timely information.

The company originally planned its GM ACCESS system to distribute new product information quickly to all the stores, but soon expanded its capability to far more functions. Using satellite technology, the company now also furnishes information about cars, such as options, availability, pricing, service information, recalls, and marketing programs. An extension accessible by consumers allows them to view parts of the information as well, and dealers and consumers are increasingly using the system to communicate with each other instead of depending on personal meetings, telephone calls, and postal mail.[17]

➤ Managing Sales for a Brewery

London-based Guinness Europe is part of Guinness Brewing Worldwide, one of the world's largest brewers. Wanting to move swiftly into Eastern Europe while also penetrating the continent's premium drinks markets, Guinness Europe needed a new approach to sales and distribution. It set up a new warehouse in the Netherlands and reorganized its distribution channels.

To support this new strategy, the company needed an information system that would simplify operations and improve turnaround times, while allowing better forecasting, inventory planning, and stock management. The company recognized the benefits of linking its production, sales, and delivery operations to financial management and accounting.

Guinness realized that integrated business software would be the best solution, since the company's competitiveness in Europe hinged on being able to exchange information efficiently with its own offices and other organizations such as shipping agents. After looking at various packages, it chose SAP's R/3 system, opting for the modules for forecasting, processing of sales and purchase orders, accounting, invoicing, MRP, warehouse management, and sales analysis. With help from SAP partners, Guinness implemented these modules on Digital Equipment Corporation computers in seven months.

The combination of R/3 and electronic data interchange (EDI) has enabled Guinness to exchange information effectively using automated messaging and real-time sharing of stock and delivery data. It can also communicate with other companies using their own systems.[18]

■ TECHNOLOGY IN RETAILING

If technology has been slow to impact some parts of the sales process, it's had an enormous impact in a major sales channel: retailing. From the security systems designed to beep when a teenager shoplifts a treasured CD to automated scanners at the supermarket, low-margin retailers have fervently embraced technology as a way to sweeten their profits.

Bar codes are an example of a low-cost innovation that required heavy commitment but now returns tremendous benefits. Virtually every item sold in supermarkets except fresh produce has a bar code that identifies it at checkout. The scanner gives the ID code to a computer that checks the price and enters it in the register. At the same time, the computer adjusts inventory, provides reports on sales and promotional activity, and, in some cases, automatically orders a replacement. This, plus a trend toward just-in-time delivery, reduces inventory as well as increasing the chances that products will be in stock when a customer wants them. Stores save time at checkout and may be able to forgo individual product pricing in favor of shelf tags.

The ubiquity of these scanners has encouraged suppliers to package and brand fresh produce like spinach and leads to industry jokes about genetically engineering tomatoes to include their bar codes. Consumers first fought the omission of prices on items and worried about incorrect prices, but that concern subsided as they enjoyed faster checkout and receipts with details about items and pricing.

Other retailers have embraced bar codes as enthusiastically as supermarkets. Woe to the consumer who accidentally grabs a bolt without a code at Home Depot. It seems as though the whole store shuts down while the staff frantically looks for the number, because the clerk can't sell a part without a stock code.

Department stores are a little less structured, but watch the clerk enter a seemingly endless stream of numbers when the laser reader can't decode the bar code, and you wonder how businesses existed before the technology did.

Ironically, bar codes almost didn't become as important as they are. Years ago, many suppliers hoped that optical character readers (OCRs) that could read almost-conventional text would become the standard. Too many errors and relatively expensive equipment doomed them. Now suppliers include the bar codes on product labels, so there's no longer much reason to consider OCRs. Even the postal service, with its expensive OCR equipment, uses it just to apply bar codes to envelopes and gives discounts to volume mailers if they'll include the codes on mail.

Of course, consumers have their own versions of bar codes—the magnetic stripes on credit and debit cards. Gas stations have gone from full-service to self-service, and now to no-service stations. At many stations, you don't interact with a human at all; you simply insert a credit card into the gas pump to start your transaction, while a person hides in a bulletproof cage dispensing cigarettes and lottery tickets, and unlocking the pumps for the few drivers without credit or debit cards.

Use of this automation has led to tremendous savings in labor costs for retailers, and much of that has actually been passed along to the consumers. Giant warehouse stores may make you bag your own groceries—a few even make you scan your purchases—but the trade-off is lower prices.

Still in the experimental stage are kiosks that provide store locations or other information, but customers still seem to need signs and maps—or humans. That may change as consumers become used to these capable automated helpers, just as they learned to accept automated bank tellers, but, so far, that hasn't happened.

➤ Data Warehousing Has a Big Impact in Retailing

Some of the biggest impact of technology in retailing has been completely invisible to consumers. Optimizing inventory has had a big impact on operations and profits at retailers as diverse as Target discount stores and the specialty retailer, Victoria's Secret.

At Victoria's Secret, a division of The Limited that sells women's intimate apparel, a survey of customers disclosed that their biggest frustration was not finding the right size or color in stock. Analysis revealed the surprising fact that stores in different areas had significantly different buying patterns. Unfortunately, the company's inventory policy was based on averages: If a store did twice the company's average volume, it got twice the inventory in the same portions, notably colors and sizes.

To solve the problem, and gain other benefits as well, the company created a sophisticated data warehouse of detailed information about sales of each item at each store. This data warehouse, a concept that has become an important issue among information technology users, stores the data such that it is easy to access in many different ways by nontechnical users, including marketing experts who analyze trends and forecast future patterns.

Implementing the warehouse turned out to be a huge job. It required very large database (VLDB) technology running on a Tandem Himalaya server and a NonStop SQL database, plus decision support software, DSS-Agent from MicroStrategy, which allowed users to analyze the data in many ways. To make it even easier for users, however, it is presented to them using familiar browser technology on the store's internal intranet.

The system has served its purpose, as current surveys indicate that finding the right size and color is no longer a major issue. The system also allows the stores to maintain lean inventory, assuring that seasonal items aren't left over. The system allows home office marketers to analyze the impact of promotions and product displays in stores without calling on busy information departments. Another benefit is the ability to instantly spot poor performance by stores and allow correction, as when some stores aren't taking advantage of popular promotions.[19, 20]

➤ Data Warehousing at Dayton Hudson

Victoria's Secret found the data warehouse concept ideal for improving the performance of its stores and, on a much larger scale, so did Dayton Hudson, parent of three different types of retailers: Target upscale discount stores, Mervyn's moderate-priced family department stores, and conventional department stores Marshall Field, Dayton's, and Hudson's. Operating almost 1,100 locations in 39 states, with a vast array of separate stock-keeping units (SKUs), DH has more than 200,000 employees and revenue over $25 billion per year.

Though DH's divisions differ significantly in inventory and operation, the corporation decided to establish one information system to serve all the stores. The company regarded its inventory management information system as more than housekeeping. It regards it as a competitive advantage for the firm. Like Victoria's Secret, DH ended up with a large Tandem system—this one with 32 processors and 1.8 terabytes (1,800,000 megabytes) of disk storage plus room to grow. DH, however, developed its own front ends instead of using off-the-shelf products. One is InfoRetriever, a general-purpose analytical processing report engine; another is In-Stocks, a "fuel gauge" agent that monitors inventory performance and issues alerts when stock gets too low.

Using this system, retail managers can check out hunches about merchandising trends, monitor fashion trends in different outlets and locations, and track sales by time, item, and location. It also helps in pricing products, negotiating with vendors, and evaluating promotions. One advantage of the system is that, once managers learn to use it, they can conduct new analyses—profitability, for example—without learning a completely new application. The company maintains 18 months of weekly sales and inventory data at the item and location level, allowing it to track seasonal patterns, trends, and effectiveness of advertising and other promotion.[21]

Like Victoria's Secret and Dayton Hudson, many other companies are turning to data warehouses, but it's a complex, expensive process that hasn't always succeeded. According to Computer Systems Corporation, many of these operations haven't paid off, but that is likely to change as companies home in on what they are trying to accomplish and gain experience.

Retailers use technology in many other ways, too. Aside from inventory management and research, one of the most interesting applications is store layout. At the Gap, for example, in-house architectural staff use 100 Sun workstations to design new stores to join the 1,800 Gap, BabyGap, GapKids, Banana Republic, and Old Navy stores. Using virtual reality, the staff can "walk" through the stores to check layout before completing designs. The technique is so appealing that the Gap has even posted a 3-D version of one of its stores on its Web site, allowing viewers to tour the store.[22]

➤ Leading Winery Uses Research Tools

Just 15 years ago, Sutter Home Winery of St. Helena, California, was a small company in the Napa Valley, whose wines were available only at the winery and in local stores. Today, the label is sold in every major market in the United States—it is the fourth-largest U.S. winery in total sales.

The company has successfully managed its phenomenal growth, and one of the tools that has helped it do this is Cognos's PowerPlay analysis software. As Bob Meanza, Sutter Home's director of sales administration explains, the alcoholic beverage business is a complex business to manage. Each state maintains regulatory power of alcoholic beverages, which means wineries cannot sell directly to stores or restaurants. All out-of-state sales must be made through each state's distributors. As a result, Sutter Home must compile a picture of its sales, profits, and distribution patterns by accumulating information from over 150 different distributors.

"We know how many cases we ship to distributors. But the real information we need is what the distributors actually sell, where they sell it, and how rapidly they sell it," said Meanza. "So we rely on our distributors to provide us with sales information that, once accumulated here at our headquarters, is distributed to our sales people across the country so they can work more effectively with their distributors."

Until recently, Sutter Home's mainframe-based information was provided to salespeople through standard printed reports. Sales managers could see columns of data, but they couldn't work with it dynamically. In order to conduct analysis, sales managers would often reenter the same data into multiple spreadsheets.

When Sutter Home set out to automate its sales force, the company's goal was to provide a tool that would help salespeople keep up with the company's rapid growth and success. The first step was to arm every salesperson with a laptop computer. But if that was all they had done, notes Meanza, "we would have been giving them nothing more than very expensive calculators." So the company evaluated a wide range of software, including decision support tools.

"We needed a flexible tool that could handle the complexities of our business. PowerPlay allows us to look at information in many different dimensions—by region, by state, by product, by distributor, whatever we need," says Meanza. Support and training were also important factors to consider. And Cognos's full education and services ensured that Sutter Home had its applications up and running in a relatively short time.

Today, Sutter Home's salespeople use PowerPlay to analyze the performance of various wines and other products in the Sutter Home portfolio, plan sales strategies, and set sales goals that both Sutter Home and the distributor can agree upon.

"We've always been true believers that if you can't measure it, you can't manage it," says Roger Trinchero, president of Sutter Home. "In the past, measuring our sales performance has been very laborious. Now with PowerPlay it is very easy for our people to zero in on any segment of the sales arena and get substantial information they can sink their teeth into."

Now that PowerPlay has proven so successful, Sutter Home is planning to expand its use of Cognos Business Intelligence tools, including Impromptu.[23]

■ DISTRIBUTION AND WAREHOUSING: THE CINDERELLAS OF BUSINESS

Distribution and warehousing are two of the least glamorous aspects of business, but poor planning in these areas can negate all the improvements a company has instituted elsewhere in its business. Discount chain Price Chopper can attest to the importance of the link between the warehouse and other business functions. Before Price Chopper implemented a new warehouse management system (WMS) in its 405,000-square-foot dry grocery facility in Schenectady, New York, customer billing deadlines hindered what the warehouse could send out.

The warehouse has 1.5 million cubic feet of storage and 36,000 pallet positions. It ships more than 30 million cases out of the facility annually, a number growing about 18 percent a year. Its average inventory is 1.1 million cases—7,800 active SKUs. It's a 7-by-24 operation.

Under the old warehouse billing system, all Price Chopper store orders were processed daily at 2 A.M. While the orders were built at various times during the next day, no new items could be added to the shipments because the billing process had already been finalized.

The new WMS from Dallas Systems allows more time to incorporate last-minute changes to outgoing orders. This reduces out-of-stock situations. Warehouse crews can now incorporate additional incoming product shipments into any store order not yet assembled. This means they'll bill and assemble fewer orders at one time, allowing many other store orders not yet assembled to receive new items as late as noon.

Another result of this installation is Price Chopper's new ability to use wireless computers on lift trucks, laying the groundwork for increased use of pallet bar coding and cross-docking.

Price Chopper's WMS requirements included the ability to see store order levels, by line item, and to make modifications to orders. The package also had to be compatible with the company's transportation routing and scheduling system, and it had to be able to accommodate any technological enhancement that came along.[24]

Transportation is also key to fast-growing Tyson Foods. For Tyson, advanced transportation planning has become more than a luxury. Tyson Foods supplies chicken, pork, beef, seafood, and tortilla products to 90 of the top 100 food-away-from-home eating establishments, from McDonald's and Taco Bell to white-tablecloth restaurants. Kroger, Safeway, and H.E.B. Grocery are among the company's largest retail grocery customers. Between retail and restaurant sales, more than 30 million people eat Tyson products each week—in the United States, Mexico, Canada, China, Great Britain, and Russia—making delivering the right product to the right place at the right time a considerable challenge.

The company's growth rate hasn't made the challenge any easier. Tyson Foods has grown from a relatively small $1 billion business in 1985 to $5 billion today, requiring many changes throughout the organization. In the area of transportation planning, change is particularly critical for two reasons: a need to improve the ability to meet growing customer demands and the need to meet the requirements of an industry initiative called Efficient Consumer Response (ECR).

Although Tyson is currently able to meet customer service demands and fill product orders, the company licensed Manugistics Transportation Planning to help Tyson ensure that customers' needs will continue to be met in the future. Tyson will also be counting on Manugistics to assist with reengineering the transportation planning process as a whole.

Tyson Foods expects that Manugistics will play a large role in ensuring quality distribution services to all of its customers. Other expected benefits include more efficient use of carriers, improved order combinations, faster disposition of loads and drops, and better routing. Using its current processes and technology, Tyson's transportation department was at its limit. Transportation planning was accomplished manually by people using scratch pads. New orders—both paper and EDI—were combined into truckload units and booked with carriers. Then products were manually assigned to refrigerated trucks and railcars or to metal containers that go to Tyson distribution centers.

When the winter holidays arrive each year, accommodating the influx of holiday orders adds to the challenge. Last-minute orders are routinely piled on top of already heavy volume. The process is labor intensive, people intensive, and time intensive. And Tyson is ready for change. Its transportation and distribution planning processes had remained static. It had been using the same systems to try to distribute five billion dollars' worth of product as with only one-fifth that amount.[25]

ECR is also prompting change, as the grocery industry seeks out innovative ways to make ECR feasible and profitable for retailers and manufacturers alike. An emerging initiative to bring better value to the consumer by rethinking the grocery supply chain, ECR's main goal is to form paperless links between multiple organizations to more efficiently and accurately match product supply and demand.

The push to move forward with ECR has begun to impact all members of the grocery supply chain, and for Tyson the pressure is already starting to come from some of its biggest retail customers. Kroger, Safeway, and H.E.B.—which also use Manugistics for transportation planning—are all in the midst of carrying out ECR initiatives.

Liz Claiborne is one of the nation's largest makers of fashion apparel and accessories, with annual sales of more than two billion dollars. To improve its distribution, Liz Claiborne chose to install DM Plus distribution and traffic management software from McHugh Freeman in Waukesha, Wisconsin, at five of its distribution centers throughout the United States. The first site was its 450,000-square-foot center in North Bergen, New Jersey.

The software is a real-time, wireless-based open systems software package for managing raw materials, components, and finished goods. Customers use table-driven parameters and attributes, user-defined fields, and unique processing provisions to achieve their specific objectives. At Claiborne, the system runs on a Hewlett-Packard HP 9000 server integrated with specialized wireless handheld computers from Norand. The system manages inbound and outbound appointment scheduling, yard management, shipping/receiving, cross-docking, and order management.[26]

DM Plus also integrates with TRACS, a real-time transportation management system for shippers in the manufacturing, supply logistics, warehousing, and distribution industries. Weseley Software Development Corporation is part of McHugh Freeman, allowing the companies to develop one of the industry's first comprehensive warehouse/transportation management (WTM) solutions.

■ TRACKING HAZARDOUS MATERIALS

All the programs discussed up to now have been positive ones, designed to improve the way companies operate their sales and distribution. However, some programs are developed to solve specific problems instead.

One relatively new requirement of the distribution process is complying with government regulations. An example is shipments of hazardous materials. Even if they aren't particularly hazardous, in fact, all chemical shipments must comply with numerous national and international regulations. The regulations, issued by the U.S. Department of Transportation, the International Maritime Organization, and the International Air Transport Authority, plus standards required by the Occupational Safety and Health Act and American National Standards Institute, must accompany shipments to describe the contents and hazards they might present. Unfortunately, these regulations aren't necessarily consistent, so a package might require many labels, and the penalties can be severe if they omit even seemingly minor words.

In the past, this information was prepared manually, a daunting task given the many materials involved, the number of regulations, and the constant changes. Just as significant, there were not necessarily standards and tests on all these materials.

To simplify the task, Air Products & Chemicals developed ChemReg, a knowledge-based computer system for automatically generating the shipping descriptions and Material Safety Data Sheets. It consists of three knowledge bases, a large relational database of information about products, and an on-line system that can be accessed throughout North America and Europe.

To address the problem of new materials, it includes a process it calls *similar product logic* to find the most appropriate matches and add their warnings to any labels. Now, the system generates the labels in a few minutes, compared to the two to four person-days of time required if the process were manual. In addition, the company estimates that it would cost $1 million to test all products for compliance, plus $1 million per year in additional costs. It also allows the company to avoid potential fines and civil liability while providing the information needed to minimize potential injury to its own workers and other people.[27]

■ TECHNOLOGY IN SELLING

Both tiny and huge efforts have paid off for companies trying to improve sales productivity, but comprehensive sales automation is still far from complete at most companies. Inexpensive pagers and contact management software have significantly improved the productivity of individual sales reps, while complex inventory management and sales force automation have paid off for firms willing to reengineer functions and even their whole companies. It's clear, however, that simply trying to impose sophisticated systems that require people to work in new ways rarely succeeds by itself. It often takes a new organization or a dedicated effort to make such systems work. Industry is littered with abandoned sales force automation projects, yet companies that have succeeded often reap dramatic results.

Unlike direct sales, retailing has already been transformed by technology, from the universal use of bar codes for collecting and tracking data to sophisticated inventory management schemes using data warehousing. The most exciting potential development in sales, however, is the Internet. As in marketing, it provides a direct channel to customers, opening new ways of selling and new financial models. While few observers expect it to replace the corner store, it will likely revolutionize some businesses. It's already impacting book selling and travel reservations, and wide availability of home listings is likely to put severe pressure on the real estate market.

"The Internet will create winners and losers in every industry. It's just a question of when it gets to that stage in each of the industries," says Louis Gerstner, CEO of IBM. "One question is the role of middlemen, insurance agents, travels agents, to name a few. Where is their future in an electronic world?"[28]

Chapter 6

When Does the Sale End and Support Begin?

■ INTRODUCTION

Supporting customers obviously makes a lot of sense; happy customers come back to buy more products and services, and they recommend helpful firms to their friends. And even if they have one bad experience, satisfied long-term customers tend to give a company the benefit of the doubt and another chance.

Though not all companies act like it, customer support is an extension of their sales process, one that can have substantial impact on their bottom lines by reducing long-term sales costs and mining new revenue opportunities. For all businesses, building strong relationships with customers is a strong trend and a winning strategy.

Most service businesses—brokerage firms, banks, and even physicians—intuitively recognize this. They know that the only real difference between sales and support functions is whether the customer is already committed—or at least has already spent some money. Every call from a customer is a sales call for these companies, even though too many of their representatives don't think that way.

The situation is less obvious for companies that sell discrete items, whether they be computers or real estate. Supporting customers who have already bought something has been considered an obligation at best, a nuisance at worst, by many companies and their employees. There is often a sense that time spent helping those who have already spent money is wasted and doesn't return any immediate revenue. Fortunately, enlightenment has helped dispel that perception, while technology, particularly communications technology, is making support less of a burden and even a big advantage. Companies can use technology to partially automate the process of supporting customers, reducing support costs—and making it easier to charge customers for the support they receive.

Many computer and software companies already expect customers to pay for support after a limited initial period, although this is an accepted concept for businesses that consumers resist. Fortunately, the rise of 900 telephone numbers for automatic billing and the increasing comfort most people have supplying credit card data over the phone or computer network make the process more and more common. Eventually, consumer-oriented companies may emulate those that sell to businesses and receive substantial revenue from the support they once gave away. Consumers are already used to paying for software upgrades that in some cases provide as much fixing as innovation. The increased use of technology to organize and monitor support services also allows management to better track support costs and revenues. For many companies, support revenue is "found" money. Charging for something you once gave away is sure to be highly leveraged—more so if you can develop other new services to sell at the same time.

■ THE BEST SUPPORT: REQUIRING NONE

It is no surprise that some of the first companies to take advantage of technology to support customers were computer and software firms. Ironically, some of the most important steps in this process were to help eliminate the need for support: Today's products are designed to be easy to use. Whereas at one time users had to be technically competent to install and use a computer and its software, much less add peripherals and accessories, today most computers are "plug and play," often even including color-coded plugs and cables. The software, as well as vexing peripherals like modems and sound-boards, are typically installed by the manufacturer, and, increasingly, Windows-based personal computers can recognize and install additions without problems—just as Apple Macintoshes always have!

Good software is designed to be intuitive once you learn a few steps—steps that may seem obvious to anyone who's used a computer until you hear the stories of a tyro trying to use his mouse as a foot pedal or his CD-ROM tray as a cup holder, or searching in vain for the "any" key when a program tells him to press any key.

Programs are looking more and more alike; while once only an experienced legal secretary could master WordPerfect with its obscure key combinations that allowed the person to format a complex page, now anyone can simply point and click to create even a complex interactive page for the World Wide Web.

Most users, in fact, consider a program defective if they have to consult the instruction book to use it. That's fortunate, because as the programs added features, the instruction books became so bloated that now many suppliers no longer supply a printed instruction book. They provide it only in electronic form on a CD-ROM or, even worse, as on-line help screens. Some of these help systems are amazingly complete, but many are confusing. With some, you can only search for a topic, rather than being able to look it up in a logical table of contents or directory, frustrating the user who may not know that *font* is the proper name for what most people sloppily call "letters" or "type."

■ CALLING THE COMPANY

Of course, sometimes users have to call for help, and not only from technical companies. Even buyers of low-tech products such as oatmeal and barbecue grills often need help, whether to report a roach in the cereal or to seek a missing bolt. In almost every case, these customers—as much as someone looking for assistance with an obscure computer problem—will encounter voice mail. In many cases, the automated voice of a voice-mail tree represents the tip of an elaborate call center.

Customer support, like sales, almost always requires voice as well as data. In this, it is unlike many other applications of technology to improve productivity. Because most customer interface is on the telephone, but a lot of information is available in computers, effective support often requires strong integration between telephones and computers. Partly for that reason, customer support has challenged digitally trained computer and software developers who regard telephone systems as archaic voodoo. It has consequently required more time to implement productive telephone-computer integration than such seemingly more complex functions as accounting, engineering, database management, and even manufacturing. It also means that, in many support applications, the heart of the system is a computerized telephone call center—an elaborate voice processing system, not a traditional data processing computer.

This has been the greatest contribution technology has made to customer support. The sophisticated telephone-computer integration of call centers allows efficient processing and monitoring of customer calls. In the future, however, the Internet will play an increasingly important role.

■ THE INTERNET: PERFECT SOLUTION FOR SERVICE

Many software and hardware buyers turn first to on-line support sites for troubleshooting, for answers to questions about how to use products, and for free software updates, peripheral drivers, and accessory and utility software that once were available only by mail, often for an annoying small but inconvenient price ("to cover shipping and handling," as the phrase goes).

These systems can pay off very quickly for vendors. Hewlett-Packard, for example, is currently saving over 90 percent of its costs by providing drivers and updates from its Electronic Support Center on the Web. That's $8 million per month in costs, and the customer gets immediate access 24 hours a day rather than having to wait for disks to be sent through the mail.[1]

Some observers even suspect use of the Internet will be fueled by the well-acknowledged male preference to look something up rather than ask for help, whether that's a learned or a gender-linked preference!

According to a report by Advanced Manufacturing Research, 82 percent of large manufacturing companies expect to use the Internet to communicate with their customers in 1998, up 11 percent from the year before. Thirty percent already allow transactions over the Internet.

The Internet is a convenient way to order accessories and other parts unlikely to be stocked by retailers. As nontechnical customers of low-tech products go on-line and become more comfortable with the Internet, many will find it easier than calling a company, navigating through a voice-mail maze (often to be put on hold), then having to carefully spell their address and provide various other information. A survey by Advanced Manufacturing Research (see Figure 6-1) found that most corporations expect to communicate with customers over the Internet.

The Internet will almost surely become the preferred route for questions and add-on orders for these customers. That won't diminish the role of the call center, however. More likely, the responsibility for all support operations will be centralized and integrated, eventually allowing data, fax, and voice interactions on the same call. For now, however, most users and suppliers alike depend on voice for the customer interface.

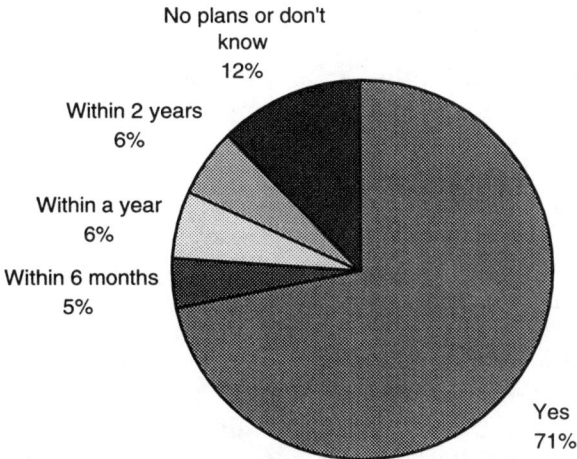

No plans or don't
know
12%

Within 2 years
6%

Within a year
6%

Within 6 months
5%

Yes
71%

Figure 6-1. A survey by Advanced Manufacturing Research found that most corporations are or expect to communicate with their customers over the Internet.

■ RESCUE FROM VOICE-MAIL HELL

Damned as living hell by most users, complex voice messaging systems may seem like paradise to companies that have used them to trim both customer support representatives and the intensive training they undergo. Even users, though frustrated at times as they navigate a voice-mail labyrinth trying to find the right person or solution, will admit that voice messaging is often an efficient mechanism for solving a problem.

Voice messaging, of course, is no longer just voice mail. Today, companies can integrate voice messaging with computers and E-mail, reading written messages to callers automatically, sending them faxes, or converting spoken words into commands or E-mail messages. Most commonly, however, these systems still depend on pushing telephone buttons in response to automated words. Unfortunately, there's never been any standardization, so on some systems, pushing "0" always connects the caller to a human operator, whereas others require a "#" or "*."

➤ Your Friendly Local Cable System

Perhaps no application better symbolizes voice-mail hell to users than calling the local cable TV company to start service, clarify billing, or order special programs. Not surprisingly, cable systems have become big users of voice messaging. One example is Cox Cable, which provides cable television service to almost 90,000 subscribers in the Spokane, Washington, area. Both Spokane and the cable service have been growing rapidly, increasing the number of subscribers by almost 40 percent in the last five years.

By 1994, its six-year-old voice-mail system was showing its age, offering only basic voice-mail features but none of the modern productivity tools that have made such systems invaluable. Cox installed an advanced Octel voice processing system that integrates voice mail with PCs on a local area network. This allows support representatives to access voice and fax messages from their PCs. Using the PC screen, the reps can see such information as type and length of message, priority, time received, and sender name or number, then click to listen to the message on their telephone receivers, using on-screen commands to pause, repeat, and fast-forward messages. It's not yet practical to actually convert voice to text, but that will come one day; in practice, voice is probably adequate for this application. Users can also add written notes to the messages and forward them to others, or they can categorize the messages.

About 65 people, including both cable and advertising sales representatives, use the Cox system. The sales reps, in particular, find the system vital. They spend most of their time outside the office on calls, so many of the calls they receive become voice messages. With a simpler system, the calls could be processed only in the order in which they were received; now such systems can identify priority messages, store saved messages by advertiser, and locate saved messages immediately by caller. Cox estimates the reps save between five minutes and an hour a day using the system, which reduces their stress immeasurably. The system also incorporates similar capability for fax, a growing medium of communication for the reps and even for homeowners.[2]

➤ Supporting Customers at US West

A different tactic used to improve customer support is improved service for the representatives who deal with the public. US West is the Denver-based telephone company that serves the mountain states. People who called it for installations and changes used to be told that starting new service took approximately 48 hours, whether it actually might have taken one hour or five days. The company had no simple mechanism that allowed a better estimate. Not surprisingly, the timing was often off. Even if service had been sooner, that was often a problem for customers, who didn't know when to expect a technician to arrive. More commonly, the length of the wait was increasing as more residents moved into the booming area and ordered extra lines for teenagers, home offices, computers, and fax machines.

Ironically, the information was available within US West's massive legacy computer system, but there was no practical way to get that information to the desks of the service reps while they were on the phone with customers.

The phone company took what is becoming an increasingly popular route to solving its problem. It developed an application that used a familiar Web browser at the representative's workstation to access the mainframe over a local area network. The browser simply launches a search engine that finds the data required, then gives a realistic estimate of when the service will actually start.

US West Vice President and CIO David Laube says using this approach allowed the company to provide its 6,000 service reps with this information in only eight weeks' time—after ten years of waiting to find a good solution. He also claims that, beyond the happy customers, the application saved $10 million for the company in its first year.[3]

➤ Waiting for the Limo

Telephone installers aren't the only people requiring reservations. High up on the list of life stresses is waiting for a reserved taxi or limousine, particularly when you're catching an important flight— or getting married. It's an equal headache for limo operators, who must juggle their fleets for maximum utilization without overlooking customers.

A-1 Limousine, the fourth-largest limousine company in the United States, serves customers in New York, New Jersey, and Connecticut from its headquarters in Princeton, with 250 limos, stretch limos, vans, and sedans. It had been managing reservations using an older database system since 1991 and, over time, its system had grown increasingly large and cumbersome, often producing errors that resulted in missed pickups, unhappy customers, and resultant refunds that impacted profitability.

On top of that, the company was attempting to grow by taking advantage of new technologies and approaches to selling its services. For one, it intended to allow customers to interact and make reservations over the Internet. For another, it intended to open remote sites with sales representatives and drivers to allow quicker response and greater revenue. All these factors encouraged the company to reevaluate its existing programs.

The firm chose Computer Automation's OpenIngres to create A-1's reservation, dispatch, and sales support system, the brains of the company. The system connects 14 reservation agents, 14 dispatchers, 3 salespersons, senior management, and, indirectly, 250 limousines. Running on IBM RS/6000 under the AIX operating system, the product has demonstrated 30 to 50 percent better response times and far greater accuracy than the earlier system. The success has encouraged the firm to plan installation of remote terminals in each limo, further automating the process and providing better management of the expensive assets.[4]

In addition to customer support, technology is assisting corporations to service their products as well. One application is scheduling service visits, a problem encountered by appliance repair technicians, as well as cable TV and telephone installers. Another example is built-in or on-line diagnosis of problems, an obvious step for computer and communications equipment, but one that is being applied to products as diverse as copy machines and commercial freezers, too. Often, technicians at a home office can not only diagnose the problem, but sometimes fix it on-line or at least tell the user of the equipment what to do. Even if it only allows technicians to make sure they have the proper replacement parts, this step can save significant time and travel. This is especially true as equipment is built in easily replaceable modules that can be swapped rather than requiring a wait for repairs. This is increasingly common for automobiles, as well. Sophisticated on-board diagnostic systems, often requiring dealer equipment, quickly pinpoint problems for immediate—if expensive—repair. Many companies will now ship replacement components, or even whole products, overnight without waiting to receive the broken part, reducing downtime for their customers.

■ PRODUCT DOCUMENTATION ON-LINE

One of the biggest problems in servicing any complex, constantly changing product is having up-to-date documentation. And where human life is concerned, it's an even more critical issue. Perhaps the most complex product ever built is the modern jet airliner. The maintenance manual for a McDonnell Douglas MD-11 jet contains 40,000 pages, for example. Printing, inventorying, and distributing these manuals to airline maintenance shops around the world is a major and expensive effort. Then there's the matter of updates.

Airplanes aren't standardized—as anyone who's experienced different seating configurations knows—and they aren't static. Every airline orders different arrangements and equipment, sometimes as major as engines from different manufacturers. And there are constant upgrades, some for comfort and marketing, some for safety. All these changes must be carefully documented, or chaos could result.

Like many software companies, airplane makers are turning to alternative technology to deliver service information. A CD-ROM, for example, can hold 650 megabytes of storage, enough to hold 40 million words, about 80,000 pages (though the diagrams needed consume considerable space). A CD-ROM is also inexpensive, costing as little as a dollar or so—far less than the cost of the printing and paper of a large manual.

Even CD-ROMs get outdated, however. Douglas Aircraft has taken one step further, delivering large segments of its manuals to mechanics over the Internet. Using a laptop computer, often with a wireless link, the repair personnel can access needed information inside an airplane, even if they're in Bangkok and the original lies on a server in Long Beach, California. The service also includes the latest updates, of course.[5]

One major issue with the program is security. Knowing that terrorists and others could find details valuable, Douglas has implemented multiple levels of passwords and security to ensure that only authorized personnel have access to the data. This includes encryption from Netscape Communications Corporation, supplier of the server software, passwords, and an Internet firewall that keeps intruders out. All of the company's documents are in or are being converted to the Standard Generalized Markup Language (SGML) format, the parent of the Hyper Text Markup Language (HTML) format of Internet pages. This allows them to be easily converted to be accessible on-line. The data itself is in an Oracle database running on a Hewlett-Packard HP 9000 UNIX server.

■ TURNING CUSTOMER SUPPORT INTO A CENTERPIECE

The Douglas maintenance system solves one vexing but specific problem for the airplane manufacturer. Cox Communications and US West also implemented specific systems that quickly proved cost-effective. Blue Shield of California, on the other hand, took a more strategic approach. One of California's largest health care insurance providers, Blue Shield undertook massive changes that moved its technology from back-office support centers to the centerpiece of the whole organization.

In 1994, when Blue Shield initiated the effort, the firm's management knew that it was undertaking what could be a wrenching experience. The firm was a venerable nonprofit group converting into a profit-making organization to help it adapt to massive changes in the industry and severe competition. Its internally focused staff had already undergone 25 percent cutbacks, site closures, and significant changes in their priorities as they faced a company-wide reengineering process to become competitive. In addition, Blue Shield's own information organizations were used to providing support, not leadership. They had little experience in aggressively planning and implementing what would become a critical part of the company.

Given the unstable situation of the health care industry at the time and Blue Shield's noncompetitive status, the organization decided to start from scratch. Instead of making evolutionary changes, Blue Shield took a "blank paper" approach, fundamentally reengineering its whole business process. An approach that has occurred in other industries, it was new to health care insurance.

Because dealing with customers is Blue Shield's business, the firm based its state-of-the-art information system on a communications perspective, not the computer mainframe viewpoint that many companies would have chosen. What resulted was a comprehensive call center operation that combined its existing systems—mainframe computers, desktop personal computers, local and wide area networks, communications switches and routing equipment, and customer and provider databases—with new, leading-edge technology—some so new that they weren't available in standard form from vendors. These new technologies merged the existing computer products to operate with the universal customer "workstation," the telephone. The new systems include computer-telephone integration (CTI), interactive voice response (IVR), and speech recognition.

Technical staff, aided by consultants and vendors, worked closely with the rest of Blue Shield's staff to plan and implement the system to meet new requirements. Both parts of the staff were challenged in changing the fundamental way they worked as they adapted their culture to the new order and developed their capabilities. Ultimately, the process resulted in benefits for all involved—

employees, health care providers, and customers—including more flexible access to information, lower insurance rates, more satisfying work, and greater operational effectiveness.

The primary purpose of the whole project was to improve Blue Shield's competitiveness to retain more customers, help acquire new group and individual accounts, improve relations with care providers, and control costs. As part of the first objective, retaining customers, the system was designed to improve basic customer transactions. It allows service representatives to greet in-calling customers by name, with their account information and profile on screen. It also routes customers to agents who know the care providers in their location, provide considerable after-hours access to customer information, and locate data using vague information, such as approximate dates and incomplete provider names. This latter requires the use of *fuzzy logic,* an often discussed but little implemented process that makes decisions based on vague data.

The system automates common tasks, such as calls about eligibility, co-payments, physician status, addresses, and claim status, letting the agents take calls in which they add value and require judgment, sensitivity, clarification, or special treatment. If the agents don't have to retrieve basic information, they can contribute more to the customer—and be happier. Even if customers decide to leave the automated process, information they have entered, such as account number, allows the agent to have their names and policy information ready. The system is also bilingual, accommodating the many Californians more comfortable speaking Spanish than English.

Physicians, hospitals, and their staffs benefit, too. They can access certain information automatically, allowing them to make a quick determination of patient status and eligibility. If satisfied with the process, they are more likely to refer patients to Blue Shield—as growing numbers do.

The increased efficiency of the system has allowed Blue Shield to reduce costs significantly as it adds service, allowing it to become more competitive and offer plans at lower rates. In an increasingly competitive arena, Blue Shield is maintaining and increasing market share. Aside from helping Blue Shield compete, it also reduces costs, allowing Blue Shield to offer lower rates, and permitting more members of society to have health insurance. The program has also raised the bar on service, forcing other providers to improve their customer support as well.

A year after the project was implemented, Blue Shield saw real benefits. It claims millions of dollars in cost savings and 25 percent higher service quality as measured by the national BlueCross BlueShield Association.

The company's target was that 15 percent of calls would be automated. After the first year, the figure was 11.5 percent overall and 20 percent in certain areas. The system handles the equivalent of 27 full-time agents. Yet those now at Blue Shield have been energized by the project. Though there was almost complete management turnover at the organization, many staff members have gained new skills, and all recognize that the company they work for has a new spirit and competitiveness.

The first phases of the project complete, the association is now planning future elements, including voice recognition for people who don't want to or can't use the keypad, the ability to order ID cards automatically, heath tips, the ability to secure claim forms and other documents on-line, and automated user surveys.

With the success of the customer service system, other departments within Blue Shield, such as medical management, installation and membership, and provider relations, are now clamoring for their own systems. Using the existing model should simplify future extensions into these areas.

Blue Shield's experience will likely be emulated by many other organizations in the health care field as market forces and government policy force this service *industry* to become more service-*oriented*. Not all applications in health care are for paying bills, however. Technology also helps detect and cure disease and injuries. What may not be as widely recognized is the role technology plays in tracking and monitoring health problems.[6]

■ IMPROVING HEALTH WITH COMPUTERS

Aside from the computers hidden or attached to much modern medical equipment, most computer systems in hospitals are there for financial purposes that differ little from those in many other businesses. There's been much talk, but little action, about developing systems that actually help care for patients. One place that has happened is Houston's MacGregor Medical Association (MMA), a large health maintenance operation that handles about 1 million visits a year from about 250,000 patients in almost 30 outpatient health care centers there and in San Antonio.

Into the late 1980s, its physicians depended on couriers who delivered patient record folders from a central database. For patients who needed last-minute care, and for many others, the records often weren't there, forcing physicians to order extra tests and treatment that could have been avoided.

The hospital couldn't find a system that could track patient records, so it felt forced to develop its own integrated clinical, scheduling, and billing system. Its first version, a character-based inquiry system written in COBOL for a mainframe, was introduced in 1989 as CIS—Clinical Information System. It has evolved into an intuitive graphical system using the client/server model, with doctors and other staff using Windows personal computers in their offices and treatment rooms to access data stored on a huge database on a Tandem computer system. Some physicians even view the records over the Internet or with a modem from their computers at home.

The application is intuitive: A doctor types a patient's name and identification or social security number, and a screen pops up, providing medical conditions, a record of all visits and treatment with any doctor, results of laboratory tests, and various alerts such as for drug allergies.

One reason for the success of the system is that it doesn't require physicians to key in data. Instead, after (or during) a visit, they simply dictate their findings for transcribing into the system. At some point, MMA hopes that speech recognition could eliminate the need for human transcription.

The same system is used for scheduling appointments and needed treatment such as regular blood or glaucoma tests or immunizations. It also tracks charges for billing.

The system has had a direct result on patient care. It makes it convenient for physicians to quickly screen patients—for example, finding three patients who were prescribed a medication that was recalled, or pinpointing pregnant patients who needed to be checked for certain conditions. It also allows department heads to monitor physicians, especially new ones, and offer advice if needed.

The HMO has no hard data on savings that resulted from the use of the system, but it believes that it has more than paid for the $8 million development and ongoing cost in staff reductions, shorter hospital stays, fewer premature births, less neonatal intensive care, fewer redundant tests—and fewer lawsuits. The organization is constantly adding features, such as ordering tests and checking drug interactions, and hopes to incorporate artificial intelligence that will highlight patients who need more attention.[7]

Other industries besides health care are undergoing similar searches for the balance between efficiency and customer orientation. Few have been impacted more than financial services.

■ THE NEW BANK ISN'T THERE

The whole concept of a bank is undergoing fundamental changes. Once a brick building in a local community, or at most one in a statewide chain, today's bank may be an automated teller machine (ATM)—perhaps at the local supermarket. Some depositors, often substantial ones, manage their money on-line, while other wealthy people expect personal service and won't deal with automation or other changes. Still other customers don't even use a conventional bank, but put their money in a stockbroker's money market fund or a credit union, both of which, for most purposes, act like retail banks.

Most banks are now investing heavily in new approaches and new technologies, all aimed at reducing costs and increasing profitability. The biggest costs are teller salaries and real estate, but ironically, the number of bank branches has actually remained static even though many mergers and acquisitions have resulted in banks shedding redundant branches. Under the terms of many of these mergers, the combined bank has to sell its closed branches to competitors. Many are newly chartered local banks, so few of the branches have actually been turned into convenience markets or restaurants. Many other small branches have been opened in supermarkets and other locations.

The most visible and successful of the steps banks have taken to increase automation is the ATM. Most people now use ATMs to withdraw money, whether at the side of a bank, in a supermarket, or at a convenience store. Many of these ATMs were established to save the banks money, but the banks now charge noncustomers for access. Though banks own many of these ATMs, an increasing number belong to other organizations that charge fees to use them. About half of the 33,000 ATMs and cash dispensers shipped in 1996 were to retail firms and other businesses, especially convenience stores, says Mentis Corporation, a consulting firm in Durham, North Carolina.

Unfortunately for banks, most people still won't deposit money in an ATM or conduct other transactions there. They want the security of seeing the bank accept their actual check and some type of record that shows the amount added to their account.

Some new ATMs are actually sophisticated kiosks that include scanning cameras and can display full images like checks, instead of just text, and that can perform other functions. Still primarily test beds, these kiosks appear to intimidate many people; many others are worried about the security of using one at a stand-alone location, preferring to stand in front of a teller in a bank, even if it doesn't have the marble walls and teller cages of yesterday.

In response to this reality, many banks are establishing mini-branches in supermarkets and other such places. Generally staffed by one or two humans in addition to an ATM and kiosk, these branches primarily try to sell shoppers accounts, loans, and other services. They'll also assist with traditional teller functions.

➤ Banking by Video

Chase Manhattan is one of the banks trying kiosk branches. Many of its prime customers don't go to branch banks anymore. That means the banks have to go where the customers go. For Chase Manhattan Bank, the nation's largest bank, that means that it must have trained customer service representatives when and where customers want to bank. Sometimes that may be a traditional bank branch between 9 A.M. and 3 P.M., but other times, it may be after 5 P.M. while the consumer is grocery shopping. "We need to be able to reach out to more customers," says Ed Closson, Chase vice president for self service and remote banking. "Interactive communications technology provides the bridge between Chase and our customers."

As part of its plan to expand market presence, Chase is piloting a new delivery channel: the Chase Interactive Banker. Based on videoconferencing technology, it supports a full range of consumer banking activities, allowing each customer to see and speak to a product specialist when opening accounts and applying for loans.

The bank has tested the concept at three branch locations, but expects to deploy video banks in minibranches throughout the country, both on a freestanding basis and at existing businesses like supermarkets and department stores. These minibranches equipped with ATMs and videoconference capabilities are much cheaper to operate than full-sized branches. A minibranch takes up as little as 50 square feet, versus 4,000 square feet for a traditional branch. With videoconferencing technology, one expert at a central location can service three or four branch sites at a time. That's a particularly appealing prospect as Chase eyes new financial and geographic markets where experts in areas like lending and investments may be in short supply.

The videoconferencing system Chase installed uses Intel ProShare technology and a custom graphical user interface designed by Ratio of Atlanta. Customers sitting at a PC-equipped kiosk can interact with a specialist as though they were seated at a desk across from one another. The customer experiences a largely hands-free transaction, as the screen displays information about bank products and services and the specialist enters information the customer offers by phone.

Seated in a central location, the specialist knows which kiosk the customer is using and can send Chase-branded product information sheets to the kiosk printer or provide on-screen comparisons of options. Information that cannot be verified verbally (such as driver's license identification) is scanned remotely by the specialist for viewing and archiving.

Closson says customers have reacted well to the technology, and no one seems distracted by the video. "Because of the interactive design of Ratio's application, the video becomes less important after the initial warm and fuzzy contact is established," he says.

Chase's banking specialists also seem comfortable with the technology. Although they do need to be more attuned to body language, as well as needing to look and sound professional, they otherwise require no special computer training. Chase values the technology because it allows the bank to leverage its staff experts to maximum advantage.

The first eight months the three branches were in operation, they took in more than $1 million in loan applications per location. Closson sees Chase Interactive Banker playing a major role in Chase's expansion plans. "We believe it will allow us to service people we never would have gotten to before," says Closson, adding, "The potential for worldwide access is very attractive to us."[8]

These video branches and isolated bankers are part of a larger trend to turn tellers who now constitute cost centers into sales representatives and even loan officers able to bring in profits. As long as those people are standing there, bank executives reason, they may as well be selling home equity loans or mutual funds instead of just changing $20 bills for the Laundromat. Of course, the tellers need education if they are to succeed; assisting that process is computer-based training that reduces the cost of training for a position with notorious turnover. Ironically, more and more loans are being made over the telephone, a process that requires a fairly low level of skill combined with an interactive processing package for the operator.

One of these days, in fact, people won't need to go into a bank to get change for a $20 bill. Europeans have already embraced the concept of *smart cards,* which store money that can be tapped for small transactions. But Americans remain resistant; these cards are hardly used in the United States. Yet anyone who's used them knows how convenient they are for telephone calls, mass transit, buying a cup of coffee, and a plethora of other applications. Many banks are gingerly testing the cards, but they will require massive investment in card readers by merchants and cash-downloading kiosks by banks before they can become widespread. Mentis estimates that it would cost more than $30 billion to make them ubiquitous in the United States.

Smart cards aside, banks have established all of these ATMs and minibranches, but they still haven't been able to close many of their convenience branches yet. They know that their most profitable customers—the ones who deposit the most money, borrow regularly, and use other services—want personal service. They aren't about to sit among the breakfast cereals and soda pop in a supermarket and conduct their financial affairs. Many, in fact, are attracted to private banking services and upstart banks that look like upscale offices. Some banks will even send your banker to visit you at home.

➤ Banking On-line Appeals to Prime Customers

A sizable number of good customers like to bank on-line. Using popular personal finance packages such as Intuit's Quicken or proprietary software for a specific bank, the user can maintain records, pay bills on-line, transfer funds between accounts, and perform many other banking services. Most banks now offer some type of home banking—even simply using a telephone—but haven't been very successful in capturing a large number of on-line depositors. Again, though set up to save banks money, most of these services cost depositors extra, discouraging many potential users. Other customers have simply abandoned the traditional bank and use other types of financial institutions, including a few Internet-based banks like Security First National Bank. Still, the Tower Group expects constant growth in on-line banking services. See Figure 6-2.

Many banking services still require specific access to specialized networks, but, more and more, banks are converting to the Internet. This creates a big problem for the banks, since there's decreasing differentiation among financial services available on-line.

As is often the case, it was the newcomers, not the established banks, that pioneered on-line financial services. One of these companies has not only pioneered use of advanced technology, but virtually created a new industry.

■ TECHNOLOGY CREATED DISCOUNT STOCK BUYING

When the Securities and Exchange Commission (SEC) banned fixed commissions on stock trades in 1971, Charles Schwab saw an

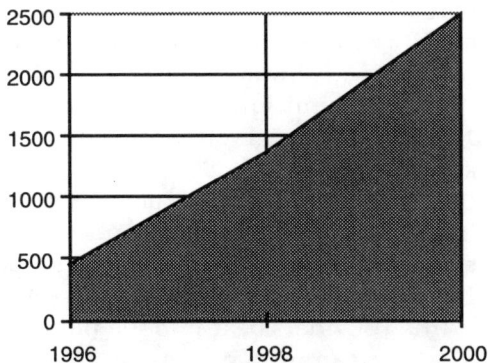

Figure 6-2. U.S. commercial banks with Internet sites. *(Source: The Tower Group.)*

opportunity. He started a discount brokerage, betting that many people simply wanted an inexpensive and convenient way to buy and sell equities. They didn't need—or want—advice from brokers. Based on that premise and aggressive use of technology and marketing innovation, Schwab's brokerage has become one of the largest in the industry at a time when there's been an explosion of interest and trading in stocks, mutual funds, options, and other securities among middle-class Americans.

The firm has consistently led the way among brokerage houses, first in discounting, then in creating a market (OneSource) that allowed customers to buy and switch among mutual funds easily— a huge advance over traditional practices.

Early in its life, Schwab bought a mainframe computer to track customers and perform trades for people who phoned their local office; then the company introduced its TeleBroker service in 1989, which provided automated quotes and trades using standard telephones. Schwab now receives 300,000 telephone calls per day, 100 million per year. The automated system handles half of these calls, with an obvious savings in staff cost. As a reward, customers who use the service get a 10 percent discount on trades. About 20 percent of Schwab's orders are placed this way.

The same year the firm introduced TeleBroker, it started brokerage services using the Genie on-line network. In 1990, the company opened its first large, centralized call centers, followed over the years by one in each U.S. time zone.

In 1993, it introduced StreetSmart, proprietary software and dial-up access that let users trade with their personal computers. Customers who only want to trade on-line and not use brokers can choose the lowest cost e.Schwab service. Schwab is also experimenting with speech recognition, offering access to stock quotes with its VoiceBroker system, which clearly is intended to allow automated spoken trading someday.

Most recently, the firm introduced Schwab Online, an Internet-based service that allows its clients to not only view research and check prices and portfolios, but execute and confirm trades instantly. It's the first such system among large brokerage firms, though other on-line providers beat it to the punch. Cowles/Simba Information reports fast growth in the market for individual on-line investor services (see Figure 6-3).

About half of Schwab's 700,000 on-line accounts come from its 4.2 million regular discount brokerage accounts. That's about half

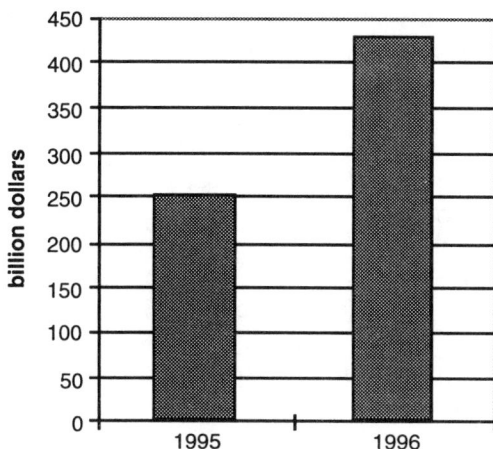

Figure 6-3. Growth of individual investor services. *(Source: Cowles/Simba Information.)*

of all on-line customers in the United States, though that number should soon explode since other brokers, such as American Express, Fidelity, and Dean Witter, have added Internet trading capability. Some other brokers, such as market leader Merrill Lynch, were slow to the party, only allowing clients to access information on-line and exchange E-mail with their brokers, but requiring them to pick up the phone and talk if they want to actually trade.

One result is that Schwab leads with younger investors who are comfortable with technology like ATMs, voice mail, pagers and cellular phones, and personal computers. Its average customer age is 47, a decade younger than that of traditional brokerage firms. These investors are also more likely to conduct their own research and make decisions rather than relying on a stockbroker or other investor advisors.

These customers are likely to continue to favor innovation, and Schwab is a legend in American business for its aggressiveness in trying new technologies—and abandoning them when they fail. Schwab was early to migrate from a mainframe architecture to distributed computing, choosing and installing 1,000 powerful Sun workstations for its brokers and managers before realizing that these UNIX computers were very expensive to buy, install, and maintain and couldn't run the software everyone wanted. Ironically, the company also discovered that UNIX software was less reliable than Windows NT, the present standard.

Now Schwab's brokers use conventional Pentium-based IBM or Compaq personal computers, with UNIX systems functioning as servers in its branch offices. The company hasn't abandoned its mainframes, however, using them for massive file management and other applications where they are most efficient.

Stock trading has two elements—customer orders and trade execution—and the company has kept customer information on the mainframes, while the session management and stock trades moved to IBM RS/6000 servers. Eventually, these relatively inexpensive servers will handle customer data as well, including Tele-Broker trades.[9-13]

➤ Full-Service Stock Broker Responds to Discounters with the Internet

One of Schwab's old-time competitors is Smith Barney, a subsidiary of Travelers Group, the investment, insurance, and consumer finance company with more than $113 billion in assets. Smith Barney's 10,500 financial consultants in 500 offices across the United States service over 5 million client accounts representing more than $450 billion in assets. The firm provides a full range of investment products, including stocks, bonds, CDs, mutual funds, IRAs, and annuities, all accessible via Smith Barney Access, its IBM-hosted Web on-line service.

As a full-service brokerage firm, Smith Barney offers personal service and advice to its clients, but it didn't have 24-hour on-line access to their records—as discount brokers do. Its clients have been requesting this access since the brokerage firm set up its marketing-oriented Web site in 1996, but the information is so personal that Smith Barney didn't want to provide access unless it was secure as well as easy to use. It now offers that service with views of brokerage positions, account transactions, selected research, and investment charts. Features include:

➤ The ability to view an instant snapshot of client brokerage positions and account value, as well as detail on recent transactions

➤ The ability to view daily prices on investments including less-published items such as unit trusts, smaller stocks, bonds, and mutual funds

➤ Access to selected research and timely investment pieces

➤ Hot links to corporate Web sites

➤ Access to Tradeline stock performance and pricing charts, which include quarterly earnings and performance compared to Standard & Poor's (S&P's)

➤ The ability to view a pie chart of asset allocation

➤ The ability to create and track a personalized stock watch list

➤ Assessment of financial needs such as retirement or college planning

To make access as simple as possible for its clients, Smith Barney offers them access through IBM's Internet Global Network, including dedicated Internet software and more than 500 local access points. Clients who use the IBM service connect 99.9 percent of the time over IBM networks on IBM-maintained systems and routers. By focusing traffic on the IBM network, Smith Barney expects IBM to monitor the network usage and enhance it as required. IBM also provides technical support for Smith Barney customers who have trouble installing software.

Future plans include allowing investors to download data into a PC package such as Quicken or Money, as well as a Smith Barney money management program.

Recent surveys indicate that Smith Barney clients liked the ease and comprehensive nature of the service, and working with the IBM network means the company can add services quickly without waiting for industry consortia to agree on standards for on-line commerce.[14]

■ TRACKING PACKAGES ELECTRONICALLY

Schwab has designed systems that accept client interface in many forms, but at least it isn't faced with the problem of express package services like Federal Express and United Parcel Service: having to get physical property around as well as moving messages.

Both FedEx and UPS have been innovators in technology, but with focus in different areas. Neither company could exist as it does without the innovative technology that each has developed. The heart of this technology is package tracking, but the firms differ in their approach.

Federal Express is well known for its hub-and-spoke network, in which airplanes rush packages from outlying locations to central sites such as its Memphis headquarters, where the packages are quickly placed on other planes heading for their destinations. But at least as important to the company's success has been its airbill tracking system, which provides a complete record of the handling, location, and destination on any package, current to minutes. With FedEx handling an average of 2.5 million packages a day—4 million at peak times—this process requires massive planning and implementation. Customers can monitor any shipment that has occurred within 15 days—adequate for a system that typically processes the package within a day or two.

The kernel of Federal Express's system is the airbill and its unique tracking number. Initially only a multipart form, the airbill is still completed manually only by individuals making small shipments.

In an attempt to automate this process, the company originally began providing volume users with complete specialized computers for generating the waybills, and for tracking shipments and expenditures. As desktop computers became more widespread, Federal Express provided software that could be used on any computer and that could dial into the FedEx corporate system. The ultimate step came with the Internet, which allows anyone to dial into the FedEx Web site and track packages, as well as extract software that generates waybills and submits billing.

On the company end, the key to tracking is the *scan.* Using handheld or mechanized bar code readers, the package is scanned every time something happens: when it is picked up, at every change of location such as from a van to a local center, center to plane, plane to hub, and ultimately when it is delivered.

The company has developed a number of proprietary devices to help in this process. Its Super-Tracker is a handheld optical bar code reader and computer used by couriers that pick up and deliver packages. It was perhaps the first palmtop computer and is representative of the special-purpose devices that have succeeded when general-purpose personal digital assistants (PDAs) have failed.

Using the tracker, the courier scans in the airbill number and also enters data for a delivery label that is printed by a small printer and attached to the package for easier reading. These trackers are also used in airports and hubs.

The Super-Tracker is used with either a mobile digitally assisted dispatch system (DADS), a radio system installed in a van, or a Smart Base, the wired version that is used in sorting facilities and hubs. Either transmits the scan number to the firm's data center in Memphis, ensuring that every step in the process is monitored.

Data about the airbills is contained in the airbill database, an IBM Information Management System (IMS) hierarchical database that actually encompasses a number of different databases containing such information as customer, shipping, and receiving locations; FedEx employee numbers; and types of scans. Each airbill can be scanned up to 40 or 50 times, so the overall amount of data collected is immense, generating more than 20 million transactions per day on two of the largest commercial mainframes in today's market.

Customers can access this data by telephone from FedEx call centers or by a multitude of on-line options. The fundamental contribution of this complex technology is that customers can now enter the process. They're not dependent on the shipper to tell them what's happening with their packages.[15]

➤ Turning a Technology Cost Center into a Profit Center

Both United Parcel Service and Federal Express have pioneered in technology, but whereas FedEx has focused on internal computer technology and depended on external communications vendors, UPS has also developed much of the communications infrastructure it uses to deliver 12 million packages daily to its 1.5 million customers in more than 200 companies. More than 250,000 employees, partners, and major customers directly use this communications network.

One step in the process has been to upgrade the private, leased X.25 packet-switched network it had previously established, replacing it with higher-speed public frame relay that provides higher performance, redundancy, higher peak capacity—and allows UPS to offer it to customers for their own communications as well. This network extends to Europe, where the company was the first customer of Global One, the consortium created by Sprint, France Telecom, and Deutsche Telekom to offer uniform telecommunications service there. UPS is also the world's largest user of cellular phones, transmitting 1.25 million data calls per day from its 66,000 vehicles. This effort was initially quite controversial, since many observers thought the analog cellular network wasn't suitable for reliable data transmission. Time proved UPS correct, though ironically it is now investigating digital technology that will probably ultimately prove superior.[16]

■ AAA FINDS A BETTER WAY

For the American Automobile Association (AAA), customer service isn't just an auxiliary part of its business, it's the whole reason the organization exists. And though AAA and its regional clubs offer many other services such as insurance, maps, and travel agencies, the real reason most people join AAA is its emergency road service.

Though it has 35 million members and invented the concept of auto clubs, AAA is facing increasing competition not only from other auto clubs but from credit card companies, cellular phone companies, and other organizations that offer alternative emergency road service and other services.

As the market became more competitive, AAA sought ways to improve service. One step it took was to migrate from its proprietary mainframe information system to a more flexible and open environment that would allow the organization to adapt to changes and add new services. AAA National chose an Informix On-Line transaction processing database server running on a Hewlett-Packard HP 9000 computer for its applications, initially moving membership processing records and emergency road service to the new system. One result: It reduced tow truck dispatch time by 33 percent. It also cut membership processing time from a week to 24 hours.

The new membership processing function gives its staff direct access to current information on-line. Agents can add services, change status, and perform other tasks using pop-up menus. Previously, they had to fill out paper forms that had to be rekeyed, then uploaded for batch processing by a mainframe. Information was never up to date, and it took a week to issue a membership card. Now it's done in 24 hours. The system has cut workload by 50 percent while improving member service.

The system also manages requests for road service. When a call comes in, the member's status and service record are automatically checked, and the dispatcher keys in the problem and vehicle location, then dispatches a tow truck and monitors its progress to make sure it arrives within 30 minutes. If no action has occurred by then, an alarm on the dispatcher's screen warns that the call is not being serviced properly.

It formerly took 11 minutes on average from the time a call was received until a tow truck was dispatched. Now it takes only 7 minutes.

The success of the system has encouraged AAA to expand it, and the organization is now adding digital maps to pinpoint locations better and is looking at a digital dispatch system that will transmit messages directly to terminals in the tow trucks.[17]

■ TECHNOLOGY IN CUSTOMER SUPPORT AND SERVICE

Technology is changing the roles and perceived value of many functions within corporations, but perhaps few have benefited as much as customer support and service. Once considered at best a nuisance, support is now seen as an opportunity, partly due to technology that has reduced its cost and made it an attractive source of revenue. It's becoming difficult to distinguish sales from support at many companies, and many firms are attempting to manage them as an integrated operation instead of as isolated functions.

At present, the technology that rules customer support and service is the telephone's sophisticated big brother, the call center. In the future, however, the Internet will shoulder a growing part of this burden—and opportunity—to solidify relations with customers while efficiently extracting additional revenue and profits from them.

Chapter

7

Smile When You Say "Personnel," Please

■ INTRODUCTION

Some of the knottiest problems facing corporations involve their personnel, but the people who have to deal with personnel problems don't like to be stuck with the "personnel" moniker. They regard themselves as experts in human resources, and if others might consider that term a bit crass, it's an accurate description of their role.

Personnel issues were one of the first to benefit from technology: Paying payroll was one of the earliest uses of computers in corporations. Now, however, the range of human resources tasks succumbing to technology is far broader. All the issues human resources must deal with are becoming computerized: finding suitable staff, introducing them to the company and its culture, training, paying, assuring fair reviews, tracking benefits, arranging pensions, even dealing with employee departures, whether voluntary or forced. Some of these issues, such as payroll and benefits, have obviously always required careful record keeping, but increased government regulation and employee lawsuits are forcing companies to also carefully monitor activities that were once quite informal.

Companies are increasingly integrating their human resources into corporate-wide systems that include modules for finance and other applications as well. The HR functions are designed to simplify operations while they reduce costs and coordinate with other corporate activities. They also reflect the growing importance of human resources functions in corporate life. Once the home of clerks that calculated deductions and paid employees, human resources departments have become key functions at most companies, as finding, motivating, training, monitoring, and keeping good employees become their biggest challenges. Human resources is becoming as critical to success as product development, marketing, or manufacturing. It's even becoming of strategic importance in areas like Silicon Valley where company loyalty is usually weak, by helping some companies retain good people.

As human resources becomes more strategic, its access to technology has improved. Once the last department to get new computers and software, it's often now the first place companies install integrated enterprise software such as PeopleSoft or SAP. This is often a precursor to rolling out integrated software to other functions such as accounting and finance or manufacturing.

These new enterprise packages automate and organize the way work flows through a corporation, in effect forcing employees to work more efficiently. In the process, they also automate functions that once required herds of clerks and other low-level employees. Human resources staff, like those in other corporate functions, are shedding administrative and clerical tasks in favor of more creative and satisfying work.

A key element of this process is self-service, which allows employees to view and modify their own records without going through personnel departments. Self-service access is a natural application for the Internet. This opens the way for corporations to turn the management of benefits over to suppliers like Healtheon and insurance carriers, allowing the companies to source out another complicated business beyond their core mission.

One indication of how internal administration is changing is the move away from paper forms to on-line equivalents. "How much paper is flowing around the company?" asks Bill Gates. "I recently had every paper form at Microsoft brought to me. And I looked at them and asked, 'Why do we have these paper forms? Everybody's got a PC. We're really connected up. We're ready to use electronic mail.' And very quickly we were able to eliminate the majority of them. Within a year, we will have gotten rid of virtually all of them."

One example was Microsoft's forms for its 401(k) plan. Gates said, "We had eight different forms—to enter and exit the plan, to change things. If an employee wanted to know what his or her status was, what his options were, he had to go and see someone. So very quickly, on top of our existing infrastructure, in a few months we've created something that has better accuracy, better accessibility, and those paper forms are simply gone."[1]

■ FINDING STAFF ON-LINE

Of course, before a company needs to worry about managing people's benefits, it has to find and hire them. Particularly in tight areas like high technology, an important new tool has joined the traditional newspaper ads, headhunters, and employee referrals—the Internet. In some respects, the Internet is an ideal mechanism for locating and screening people—at least people who use it. That, of course, includes technical people, whether computer programmers or semiconductor engineers. Financial, marketing, and sales staff are also becoming heavy Internet users, too.

The primary advantage of the Internet is its ability to post openings at sites used by its potential employees and collect response immediately. This is particularly true for companies looking for specialists. Technical companies such as Harris Electronic Systems in Melbourne, Florida, beyond the notice of most high-tech employees, and Synopsys, in the heart of Silicon Valley, equally praise such sources as the SuperSite, an Internet job-matching service for technical professionals. The SuperSite includes links to publications such as *JavaWorld, NetscapeWorld,* and *SunWorld* and their Web sites, read by 350,000 specialists in hot fields. Synopsys, for one, finds all the résumés it receives via this site worth considering, far more than those resulting from newspaper ads.[2] Résumés received over the Internet can be screened for certain key phrases such as "Java programmer" or "Internet author," a favorite tactic of recruiting departments.

The bad part of Internet recruiting is the opposite of selectivity. It's very easy for prospects to flood corporations, recruiters, and employment sites with their résumés. Depending on the Internet also obviously excludes a huge number of prospects who don't have access. Even public access to the Internet in libraries and other places is unlikely to make this avenue successful with a large segment of the population. Nevertheless, many observers predict that the Internet will soon become a major medium for employment ads. Many newspapers have established their own on-line classified operations as a backup in case the Internet impacts one of the most profitable areas of their business.

➤ Automated Résumé Scanning Speeds Review Cycle

For many firms, there's the problem of what to do with all the résumés that come over the Internet—or even over the transom. One company that faced this problem is UPS Airlines, Kentucky's largest private employer with 11,500 employees in the Louisville area. If word gets out that UPS Airlines is hiring, a blizzard of résumés blankets the company. When the local newspaper announced that UPS had 145 openings, for example, the company received 3,000 résumés in three weeks. It receives about 30,000 total per year. Screening them to see if they match job openings is a significant challenge. And it's complicated by the company's preference for hiring from within, so current employees have to be included in the search.

In the past, the company had five different employment centers with little sharing of data, a long cycle time, and difficulty tracking the status of a particular application. Now a centralized group checks the available external and internal pools using Resumix software, letting it process résumés very quickly, typically in a day.

The company soon found some new uses for the system. It helps smooth job transfer among union members, who are allowed to bid on new jobs. It also identifies those who want to change jobs, and the company can warn its managers to be prepared to replace them. The system also helps spouses of transferred employees find jobs—especially important since the company has a policy against hiring relatives. It makes their résumés available to other local companies, so they can look there first before placing an ad or hiring a recruiter.

The company hires about 12,000 part-time manual workers a year, and it is also planning to provide a kiosk where walk-in applicants can find out about openings. All in all, the system helps UPS Airlines' human resources staff focus on people, not procedures and paper.[3]

Another company that uses résumé scanning software is Texas Instruments, which has 60,000 employees, more than 100,000 résumés on hand, and 2,000 to 4,000 submitted by applicants each week. Performing even complex searches in seconds, it eliminates a former problem of "résumé hoarding," whereby recruiters keep good résumés, forwarding only those they can't use.

Now all résumés are sent to one center, where they are scanned and made available to all business units. The new process allows recruiters who visit college campuses to collect résumés for all units, not just their own. A major advantage is a dramatic cut in processing time to acknowledge and act upon applications, reducing inquiries that formerly disrupted the human resources staff. This allows them to respond quickly if they do get calls. TI is using the system in 25 U.S. locations and is expanding it into Asia, where it has numerous sites. The company is also integrating it into its PeopleSoft human resources management, automating the process of capturing data about newly hired staff members. TI is looking at providing intranet access so managers can respond to job openings on-line.[4]

At Network Equipment Technology, a similar system helped eliminate two of three positions for screening résumés, reduced long-open requisitions from 40 to 15 percent, and saved $300,000 annually in search firm expenditures.[5]

Unfortunately, the technique of scanning mailed or E-mailed résumés for the right words may backfire. Companies sometimes set such tight boundaries that the few applicants who qualify by experience or education can often command unreasonably high salaries, while equally capable people go jobless because of a few words.

■ FINDING SUITABLE STAFF ON STAFF

The problem of posting openings internally is a common one. At Ford's Car Product Development (CPD) division, internal job posting was a manual, costly, and paper-intensive process. Each group within CPD had its own process and employees were not aware of opportunities outside of their own group. Vacancies typically took up to eight weeks to fill, and occasionally some jobs were not filled at all.

Ford wanted a system that would transfer the input for jobs from human resources personnel to the hiring managers and employees, empowering them to make decisions and discover job opportunities previously unknown. The new system needed to interact with its Resumix applicant-tracking system, which Ford was using for both internal and external job candidates. It also needed to be easily available to all employees eligible to use it, at any time of day and from anywhere in the world. The telephone is an ideal input device for this situation: Employees may call in at any time and from any place.

Talisman Technologies, a consulting firm specializing in implementation of human resources technology, helped Ford redesign its requisition form, which is now published on its main-frame-based IBM PROFS E-mail. Managers call up and fill out this form on their PCs. It is then forwarded to finance and salary administration for approval, and then to human resources for entry into the Resumix system, where, nightly, an Edify voice-response system checks the server for new jobs and stores that information in a local database accessible to a caller who can apply for a position.

When employees call in, they can check the status of their résumés, check particular requisitions, hear a list or get a fax of positions, or listen to current job postings selected by location, category, and other information. Ford's program began in June 1994 with beginning nonmanagement levels, then was expanded a year later to include all 13,000 employees. It expects to expand the program to Europe, Latin America, and Asia.[6]

■ ORIENTING EMPLOYEES

Technology also helps reduce drudgery for human resources personnel and other staff members who have to process new employees. One of the most important functions of HR, particularly in this time of stringent government regulations, is orientation. Employees have to be given extensive information about conditions and benefits, and they have to acknowledge that they've received that information. One convenient way to accomplish this is a self-directing computer program or kiosk that takes an employee through all the necessary steps, ensuring that he or she understands and acknowledges all their significance. It can even be conducted in different languages, a major concern at companies that hire people with native tongues other than English—as most companies do these days, especially on both coasts and in the southwestern part of the United States.

Once employees are hired, the most obvious place technology assists personnel functions is in the prosaic preparation of payroll. It's difficult to imagine any company that hasn't automated the process of paying its employees and keeping track of the many taxes, other payments, and records required by federal, state, and local governments. Even the most basic accounting software packages such as One-Write Plus include payroll functions, and the increasing power of personal computers has made it possible to implement powerful payroll packages in inexpensive programs such as Quick Books, which has more than 80 percent of the accounting market for small businesses.

These programs have turned a nightmare of paperwork into a mere bad dream for most small companies. Easy enough for the owner to use without help, they've tremendously reduced the time it takes to accomplish an onerous task, allowing him or her (and the majority of small new businesses are started by women) more time to run the business.

■ INTEGRATED HUMAN RESOURCES PACKAGES

Above about $10 million in sales, however, most companies turn to more sophisticated human resources and payroll packages. Some are designed for specific types of businesses, while other generic software requires customization for each specific company.

Many of the vertical packages are designed for small or medium-sized companies, but enterprise-wide software vendors have created special versions for markets like government, public sector, and medical services, where there are similar but more complex needs.

One major software vendor, PeopleSoft, has developed specific packages for these markets as well as for general manufacturing, financial services, retailing, and higher education. These are integrated enterprise packages, not just stand-alone personnel applications. PeopleSoft started with HR (as can be deduced from its name), but other companies that began with financial accounting software have also moved into this market.

The big problem is no longer paying people. It's creating complete data files on employees that at the same time are secure yet accessible. There's a trend to allowing employees self-service access to update their own personal data (and link it to payroll functions) and to view their own benefits, pensions, and other information on-line. These files often contain employee photos; in the future, they will likely also contain audio and video records as well.

One company that's led the way in providing employees with access to data is, not surprisingly, PeopleSoft, the Pleasanton, California, company that is a leading supplier of human resources software. Putting its money where its mouth is, PeopleSoft, in cooperation with consultant Coopers & Lybrand, set up an extensive site on its own intranet to educate its employees about its 401(k) plan and let them plan and monitor their own accounts.

Working with the plan's trustee, Charles Schwab, Coopers & Lybrand designed a program that integrates the employee's own account information with details about the company's plan and current investment information. This goes far beyond the on-line information many companies have posted on their internal networks, which are typically just electronic versions of printed booklets.

The interactive format of the intranet (an internal Internet) is ideal for this purpose. Besides providing general information, it allows employees to focus on their own situations, rather than getting lost in things that don't apply to them. It also allows them to investigate options, such as the impact of investing different amounts of their paychecks or balancing safety versus yield in choosing self-directed investments.

One trade-off is that between providing access to information when employees need it and reminding them when its time to do something. PeopleSoft sends out quarterly statements to employees via E-mail to tell them how much they have in their accounts, then provides hyperlinks that allow them to access additional information as needed—even on-line investment data via the Internet.

For PeopleSoft, the provision of on-line pension information was partly a way to minimize HR staff. Most employee questions can be answered on-line, freeing the personnel department for other duties. The company requires only two people to handle 1,200 employees, with 30 to 50 new employees added monthly.

Obviously, the PeopleSoft staff is used to using computers. At other firms, touch-screen kiosks could serve the same purpose. Not everyone will be comfortable accessing data this way, but it wasn't that long ago that "experts" said Americans wouldn't abandon bank tellers for ATMs.

The information about 401(k) plans is a first step, but there are many other potential applications for intranets to help in human resource management. Other possibilities are benefit plan descriptions, access to forms, flexible program explanations and enrollment, directories of approved health care providers, company stock information, even virtual company meetings, particularly for geographically dispersed companies.[7]

To make it easy to implement this capability, software companies are developing specialized add-ons for human resources programs. NetDynamics, for example, offers a personnel intranet package for PeopleSoft software. This allows employees to transact the most common functions, such as corporate directories, personal information, emergency contacts, job listings and postings, benefit plans, dependent records, and direct reports.[8] Carolina Power & Light provided a human relations intranet using the package for its 6,700 employees, including job postings and salary planning. The software also allows companies to develop custom applications. Glaxo Wellcome, for example, provided a salary-planning function to its 700 managers and expects to install other intranet human resources functions as well.[9] Altogether, the use of high-tech communications tools is growing rapidly, as Figure 7-1 shows.

Stepping beyond today's most popular Internet technology, exciting multimedia presentations could do an even better job of educating employees—particularly those from the MTV generation—about subjects that seem fundamentally boring until you need them. For now, interactive computer presentations are the norm, but the Internet is sure to assume this function soon.

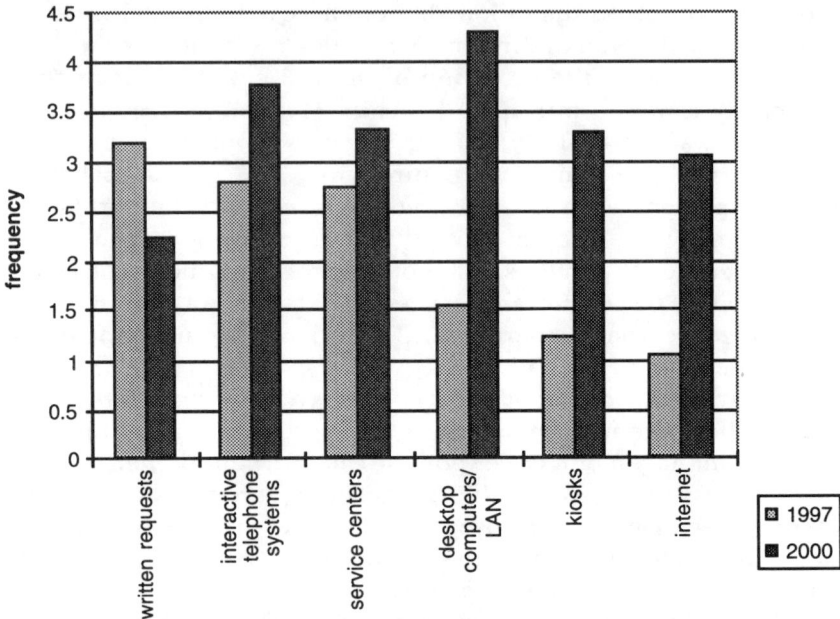

Figure 7-1. High-tech communications tools are growing rapidly in importance, according to a survey by the Hunter Group and Organizational Resource Counselors, Inc.

■ CLIENT/SERVER SERVES CLIENT NEEDS

Look closely and the amount of information companies need to store is substantial. Besides personal data, the files need to contain reviews and records of disciplinary processes. For many companies, records cover facilities in different locations, different states, and even different countries, all having specific and sometimes conflicting demands. More and more, this means that companies are adopting distributed client/server systems, whereby information is kept in central (or multiple) databases, with access, reporting, and payroll generation performed at different sites, generally interacting with a graphical front end. Client/server systems have many advantages, but they're also more complex than the earlier monolithic mainframe applications, a situation that has created many opportunities for consultants and systems analysts to prosper. Price Waterhouse's Global SAP Practice includes 2,900 trained and experienced consultants, for example.

SAP's enterprise-wide R/3 system is sold on the basis that massive improvement in performance is worth the expense and commitment to a totally new way of conducting business. Virtually always sold at the highest levels of corporations, SAP software typically requires extensive support from leading consulting firms for implementation. In reengineering itself from an old-line mainframe supplier to a client/server specialist, SAP has undertaken a remarkable transformation unmatched by any of its long-term competitors.

➤ SAP Human Resources at Motorola

One of the largest installations of SAP's R/3 in the United States has been at Motorola's Semiconductor Product Sector (SPS), a major supplier of integrated circuits, with numerous facilities around the country and around the world. With assistance from Price Waterhouse, Motorola implemented the human resources and payroll applications of the SAP R/3 enterprise client/server suite for all its 25,000 U.S.-based SPS employees and has established a foundation for managing information for its 50,000 employees worldwide.

In the implementation, Motorola SPS created a single human resources service center, allowing the company to handle more services and people, leading to increased productivity and growth. It is investigating self-service access to data over a company intranet, which would let employees make sure that their personnel data is up to date.

Before it implemented this system, Motorola SPS had a series of unrelated HR systems. The company had a payroll system and an HR master data system, each with its own database, plus separate payroll departments and record centers at sites in Arizona and Texas. After integrating its HR software in the United States, the firm is now beginning to establish a consistent master employee database based upon R/3 in its 11 sites in Europe and the Asia-Pacific region.

With the new software, Motorola is better equipped to manage its employee resources on a global level. It will quickly be able to determine where employee skills are located so the company can gather a team of the best employees to work on any given project. The HR module will also enable Motorola SPS to route personnel, procurement, and task requirements to the managers and staff worldwide that need that information.

The software required considerable adaptation to meet Motorola's needs, of course. Price Waterhouse served as the systems integrator, using the SAP R/3 Toolkit for customizing the HR and payroll applications and for managing the project. It wasn't a trivial process, because the consulting firm had to create a model of Motorola's HR processes and work with the company to decide how it should operate.

The project took 18 months, and in the two weeks after it went live, Motorola's support center received 3,000 employee calls, input 2,000 transactions, printed 2,000 employee checks, and created 22,000 direct-deposit advances with no disruption in services.

Motorola is now implementing the SAP HR/Payroll system in its Cellular Subscriber Sector, the Messaging and Information and Media Sector, the Land Mobile Products Sector, the Automotive, Energy and Components Sector, and the Cellular Networks and Space Sector. The system will ultimately serve 139,000 employees in Motorola's worldwide operations.

The next phase will add the same capability in its European and Asian-Pacific regions. At the same time, SPS will take major steps in supporting the sector with rapid implementation of key personnel development systems for recruitment, career and succession planning, and education and training management. Burger King is another global company that installed SAP's human resources management module—also for 35,000 people, but in 56 countries and obviously many sites.[10]

➤ SAP HR at Microsoft

Another leading technology company in addition to Motorola that installed the same software is Microsoft Corporation, which also implemented SAP's R/3 human resources package for both domestic and international personnel administration. Microsoft has added the human resources module to its existing R/3 system, which already ran its worldwide financial and procurement operations. The firm does business in more than 60 countries and wanted to have one personnel system integrated with finance to provide an additional level of support for managing the business.

Microsoft's central database is located on its campus in Redmond, Washington, with remote sites communicating over Microsoft's worldwide wide area network. The current system supports 2,000 users—400 concurrently—and has simplified the firm's worldwide financial systems considerably.

Microsoft saw significant benefits in implementing an enterprise solution for human resources. It replaces a dozen HR systems and 18 separate personnel-tracking databases around the world. It centralizes support while providing a localized solution for each site.

It is no surprise that the system runs on Microsoft systems software, including its Windows NT server network operating system with the Microsoft SQL Server database manager.[11]

Motorola's and Microsoft's installations were complicated by their size and many sites, but Ethyl Corporation had even more obstacles to overcome when it decided to upgrade its HR functions. Ethyl, of course, is the manufacturer that once supplied the lead that is no longer added to gasoline. That, however, was only part of its product line. It's a specialty chemical supplier with about 2,000 employees worldwide.

Like many companies, Ethyl has reorganized its business in the recent past, and during the time it was upgrading its human resources functions, the company split into five pieces while adding numerous overseas sites to its centralized records. The company started the upgrade in 1994, committing to PeopleSoft human relations, payroll, and benefits packages. As is usually the case, it hired a consulting integrator, the Hunter Group. Hunter's role grew during the process when Ethyl found its internal staff too occupied with day-to-day operations as it lost qualified personnel to the four parts of the corporation that split.

Some of the problems that were soon identified were common ones that found easy solution, but others were a bit obscure. First, Ethyl had no automated system for collecting time worked, but the software requires hourly workers to enter hours worked and rate. Complicating the issue, Hunter found different practices and rules covering various union workers at Ethyl's four U.S. plants. The consultants had to design a special time and attendance package to integrate with the basic software; fortunately, PeopleSoft, like its competitors, provides a special environment to permit this. It looks like the rest of the PeopleSoft package and operates as quickly, so it is transparent to the user.

Now many packages include functions that once had to be customized, but Hunter found that it had to write a number of special modules to add to the software. These included electronic tax reporting on a monthly and quarterly basis, U.S. Savings Bond orders to the Federal Reserve Bank for employees, the ability to download data to diskettes for off-site workers' compensation management, nightly transfer of data to Ethyl's pension administration system, internal data interface to Ethyl's executive information system, transmittal of medical and life insurance eligibility to carriers, direct payroll deposits, and transmission of 401(k) savings plan information for external management.

Overseas, Ethyl ran into an unusual request: to retain original employee ID numbers instead of assigning more convenient numbers. This simple request required the firm to adopt a special module that checked any newly assigned numbers as it filled in the gaps to make sure it didn't generate a duplicate account. The company's global nature presented additional challenges. International employees were paid by local organizations, but records had to be kept centrally. Taxes and benefits also were complicated, including the need to pay certain senior and nonlocal staff unique allowances. In addition, the company had to retain employee history for all Ethyl employees, even those who split off into the new corporations—20 years' worth with 185,000 records, a system that retains data from 150 different benefit plans and reports on 28,000 former employees under 16,000 unique job codes.

In spite of the challenges, the vendors and staff met their deadlines and, after two months running in parallel with the old systems, has since functioned as planned.[12]

➤ A Big Bang Conversion to PeopleSoft

PeopleSoft packages are designed to be added incrementally, allowing companies to bring functions up one at a time rather than forcing a complete change in the way a company operates. One company that has committed fully to PeopleSoft software is Hewlett-Packard.

In the late 1980s, HP realized that it had 40 personnel databases in the United States alone and 150 in its operations around the world. This duplication of effort inflated support costs and created many inefficiencies; the company couldn't even figure out exactly how many employees it had at any time (the number is now more than 110,000). The company decided to implement an integrated, company-wide human relations management system.

Today HP's human resources department boasts a number of programs that grew out of this move. This includes an on-line compensation reporting system that handles salary planning, an employment management system that handles staffing and job requisitions, an optical file system that stores employee records on optical media, and a worldwide employment database. With these tools, HP saves time and costs, and does a better job of assuring its employees fair and consistent treatment.

When HP decided to choose a new system, it had a number of goals. It wanted to move its payroll system off an IBM mainframe and integrate its personnel, payroll, and benefits systems. It wanted to consolidate HR activities yet maintain flexibility, and it believed that a client/server architecture was the best way to accomplish this.

After extensive investigation, HP decided that it had to either write its own software or commit to PeopleSoft—in 1990, that was the only choice for a client/server human resources package, though there are now a number of alternatives. Fortunately, HP thought the package was appropriate, but it also liked the fact that PeopleSoft provided extensive tools to allow the number-two computer company to heavily customize it.

Surprisingly, this was one of HP's first experiences with packaged software. It had always developed its own; in effect, it was a change in business processes, a first step that has been followed since. The company consolidated 40 databases into a single one, and reduced HR head count from 1 employee in 56 to 1 in 75 in the process.

HP now uses PeopleSoft HRMS (human resources management system) and payroll system with HP's own benefits system. The integrated system replaces the firm's 35-year-old mainframe payroll system. The company is also upgrading from an old HP 3000 minicomputer to a new HP 9000 computer running an Oracle database.

The list of transaction-oriented applications the group manages is impressive: PeopleSoft Human Resources, PeopleSoft Payroll and Benefits System, the employment management system, the management compensation system, the training management system, the telephone-activated benefits system, and the interna-

tional assignment application. It also maintains three information warehouses: its own PeopleBase warehouse, a worldwide
employee database, and a worldwide information database for education. One system keeps track of all open job requisitions in the
company. It also maintains résumés, with 75,000 to 100,000 on-line
for managers to view.

One major issue for HP remains its worldwide nature. The
company operates in 120 countries, and many have different processes and procedures. HP is quite decentralized, so each region
implements its own version of the software. This is one reason the
software has been so highly customized. The company would like
to standardize more of its processes to reduce this complexity.
Because of the way HP operates, its European human resources
department has worldwide responsibility for certain functions,
including a stock allocation application, a leadership development
application, and two of the hottest functions in human resources:
a Web-based job-posting application directly linked to the People-
Soft HRMS and an employee self-service application that lets
employees change personal data.[13]

➤ Managing School Buses with Technology

It's not surprising that a high-tech company like Hewlett-Packard
uses sophisticated technology to manage its human resources—it
would be more surprising if it didn't. But many low-tech businesses
do, too. One of those is Laidlaw Transit, a division of Burlington,
Ontario–based Laidlaw. It is one of the largest operators of school
buses and other transit vehicles in North America. Its 45,000
employees operate 33,800 vehicles in 43 states and six provinces,
among other efforts, transporting 1.5 million students daily.

Like HP, Laidlaw chose PeopleSoft Human Resources software.
Taking a careful approach, the company rolled out the product in
phases. It first created files on 38,000 employees in its school bus and
urban transit operations. It started with existing data in payroll files,
then added additional information from other sources. The second
step was to add training administration and health and safety modules. The third phase was to implement the benefits package.

The files are stored in a central database on an Oracle database
on a Hewlett-Packard UNIX computer in New Jersey. Managers and
schedulers access the information through the company's wide

area network using Windows-based workstations. The system's security features ensure that only authorized users have access to private information; employees can also check their files to make sure they are accurate.

The new software can track important information such as whether employees' licenses are valid and if they've completed government-mandated safety training. It even facilitates required random drug testing of drivers by automatically generating lists of eligible employees. These lists are sent to an outside company for random selection.

For Laidlaw—unlike HP—the fact that the product worked without a lot of customization was ideal. It requires little maintenance and is easy for users to operate with little training. Having an integrated source of information has also helped the company better plan salary, benefits, and career opportunities.[14]

There are many other integrated HR programs available. To serve its human resource needs, Winn-Dixie Stores selected ULTIPRO for Windows from Ultimate Software Group, a developer of human resource management systems. Winn-Dixie employs more than 135,000 persons and operates 1,181 stores in 14 states and the Bahamas. ULTIPRO for Windows replaced its mainframe-based HRMS to manage human resources, benefits, recruitment and staffing, payroll, employee self-service via Internet/Intranet, and other important HRMS functions. ULTIPRO for Windows provides Winn-Dixie with greater flexibility in cross-division reporting and significant improvement in ease of use. It will also help Winn-Dixie deal with year-2000 compliance, avoiding the need to recode and reengineer its existing mainframe system.[15]

■ SPECIALIZED SOFTWARE FOR SPECIALIZED BUSINESSES

While most large companies end up customizing human resources software for their particular needs, many medium-size firms—and some giants—find that specialized packages designed for their type of business fit the bill. One example is retail stores. There are certain unique requirements shared by all retailers—notably the problems of scheduling workers and keeping track of their time worked. One company that took this route is Abercrombie & Fitch, a specialty retail arm of The Limited. It uses two packages designed for retailers: StaffWorks and Campbell Time and Attendance, from Campbell Software. The chain chose the software for its 80 stores after a pilot test suggested it could lead to significant improvements in staffing and savings in payroll.

A&F first looked at automated labor scheduling to streamline operations in 1994. Until then, its focus had been on creating attractive stores that generated strong floor traffic, but the chain increasingly thought that its sales force could be more productive. It then initiated a program to improve staffing and customer service efficiency.

The store was the first Campbell customer to integrate the time and attendance package directly with the stores' cash registers. Combined with the StaffWorks workforce management, this eliminated the need for separate employee time clocks, which made the system even more attractive since it doesn't require stores to buy new clocks. The system will also work with time clocks or by logging on at backroom computers.[16]

That innovation aside, the timekeeping package doesn't break new ground. The heart of the Campbell innovation is its Intelligent Optimization technology that finds the best possible schedule after evaluating millions of variables such as payroll requirements, point-of-sale data, customer traffic, employee skill levels, union rules, management priorities, and other factors. Starbucks, the rapidly growing chain of coffee shops, also uses the software in its 800-plus stores.

The time-and-billing software can work with standard human resources and payroll packages by feeding information through. In addition, Campbell is integrating its front-end employee software with SAP's enterprise-wide products, including HR, accounting, and supply chain management, offering close to a fully integrated system for running a whole retail enterprise.

■ HUMAN RESOURCES FOR AN HR COMPANY

Choosing a software package designed for a specific type of business doesn't always work, however. Sometimes a company is so big, or operates so differently, that a standard program doesn't work. That was the case with Snelling International, one of the industry's largest placement agencies.

There are many packages designed for the industry's 30,000-plus placement agencies, but Snelling's needs were too big. It has more than 250 company-owned and franchise branches in five countries, and processes over 50,000 W-2 forms and 350,000 paychecks in a typical year. It also has to send reports to a thousand national, state, and local regulatory agencies.

Snelling has grown rapidly and by 1994 its five-year-old payroll package was taking a week to print out the payroll for its average workforce of 6,500 employees. It lacked the features any modern manager expects. It didn't even have an automatic interface to company financial software, it couldn't consolidate results across Snelling's different operations, and it didn't include decision support. Top management at the company wanted to choose an industry-specific package that would run on its existing IBM AS/400 minicomputer and company LAN, but none of the packages available had the capability Snelling needed.

It had identified almost 250 key features, including security, check and payroll processing, taxes, accounts receivable, and workers' compensation and HR-specific items such as insurance, Equal Employment Opportunity Commission (EEOC) requirements, tracking new hires, tracking vacations and holidays, benefits, skills and experience, job and work experience, and résumé information. One need was to track training, both to help employees progress and for required safety training.

An especially knotty problem was garnishments. Many temporary workers have multiple garnishments of their wages from different localities, even from different states, for everything from debts from car loans to child support. Sometimes the totals exceed wages, and Snelling needed software that could deal with these issues—perhaps for an employee living in one place, working in another, and with court judgments from still others. Complex, but real, situations.

Snelling discovered that PeopleSoft had the closest match for what it needed—but was still required to customize many features using a proprietary set of programs called PeopleTools, along with standard languages like COBOL and C. PeopleSoft also allows users to use standard SQL calls, providing access to its databases with industry-standard procedures. At the time, PeopleSoft didn't have standard financials, a situation that has since changed. Snelling would have bought its payroll/HR package in any case.[17]

the good guys! is another firm that went with a general-purpose package rather than one specific to its industry, retailing. The 80-store chain sells specialty consumer electronic goods, and a major effort of its adoption of new HR software was to allow access to data for different purposes without having to turn to technical experts. In the past, the database of employee information was basically inaccessible for analyses and combination with other data the chain maintained. It had multiple goals, however. It also wanted to improve processing efficiency, create business analysis tools, increase accuracy of data, eliminate redundant data entry and manipulation, and enhance management reporting. As in most cases, the effort required a hard look at the company's processes to determine what technical and nontechnical changes could be made to free staff time to better address the "human" part of human resources.

An example: The company streamlined hiring by creating pools of applicants based on their proximity to multiple potential work sites. Overall, it reduced by 20 percent the amount of time required to follow up on missing orientation data for new hires, reduced by 30 percent the amount of time to process merit increases, and saved some 4,000 hours per year in creating reports. The package selected: the SmartStream human resources package from Geac (formerly Dun & Bradstreet software).

The new program allowed the company to plan, track, and evaluate the growth of its employees—a strong emphasis. The firm offers an extensive internal training and development program, including the ability to develop custom training plans for each newly hired employee, along with management tools that help monitor the progress of the employee.

Using the open architecture of the system, the human resources package is integrated with other software used at the company. This allows staff to use other software for analysis and to develop new features, as well as transmit reports and other documents via company E-mail and develop automatic links to benefit carriers' information systems.[18]

Another company that adopted the same software is LSI Logic Corporation, a supplier of custom, high-performance semiconductors headquartered in Milpitas, California. It recently refined its human resource function with the objectives of minimizing manual processing time and costs, meeting requests for more advanced applications, and letting employees access certain data, freeing up HR staff time to focus on more strategic tasks.

LSI Logic's Human Resources Department manages benefits and payroll for more than 4,000 employees worldwide, and its mainframe system couldn't keep up with the demand for information access and management. Each time an LSI Logic employee was hired or transferred or a change was made in his or her record, HR had to route an employee profile document to members of the human resources staff, and to managers and other departments for approval and comments. Staff members had to manually deliver, fill out, approve, and pass along the document to other staff members, and HR personnel were spending time on routine administration, rather than examining organizational issues and helping to enhance LSI Logic's business. Aware of the impact the process had, human resources personnel conducted a study to determine the price tag and time required for routing employee documentation for approval. The manual process required at least seven people to spend five or more minutes reviewing employee information. This process did not factor in time spent trying to locate documents. It typically took two weeks or more for a document to travel through the approval cycle. HR employees had to deliver documents from building to building and make numerous follow-up telephone calls. Company officials realized this process needed to be streamlined.

In addition, new methods were necessary to meet the ever changing demands that human resources departments constantly face. Under the department's old record-keeping system, which was a hybrid mainframe, paper-based system, change was difficult. It was time for LSI Logic to implement a distributed operating environment, complete with user-friendly software, to accomplish the company's human resources requirements more efficiently and cost-effectively.

The department replaced its existing software with Smart-Stream Human Resources, a suite of client/server applications for managing personnel, payroll, and benefits information tasks for human resources departments. It enables employees to enjoy some self-service privileges, such as correcting employment profiles to reflect name and address changes, updating their W-4 forms, and changing beneficiaries. These changes can be performed on-line and require no filing, with easy accessibility for management review.

LSI's old mainframe system could not keep up with user demand. The company constantly found new requirements and tried to build those in, but finally couldn't add enhancements. In addition, mainframe users were growing more technologically sophisticated and were asking for graphical user interfaces with which to work. The whole world was shifting to Windows, and once users were exposed to its ease of use, they demanded it in all the applications they used.

LSI Logic's challenge was clear: Minimize manual processing time and costs, while at the same time meet user requests for more advanced applications. To address this challenge, the department launched a business process reengineering initiative. Human Resources is the first LSI Logic department to implement an early-release site test for moving from a mainframe system to a client/server computing environment. The new solutions consist of Sun Microsystems servers running the Solaris operating environment with Sybase SQL Server as the relational database management system. In addition, Compaq ProLiant servers run the Microsoft Windows NT server operating system and users were upgraded from 486 PCs to Pentium-based machines that support Microsoft's Windows 95 operating system. The department will be able to automate a range of processes, including staff management, payroll, and benefits, as well as a variety of other human resources–related functions. In addition, SmartStream will allow for more in-depth analysis and flexible reporting capabilities to Human Resources executives. The key result that Human Resources expects to deliver by final implementation is the ability to process information much more quickly with more accurate data.

By replacing its obsolete systems with SmartStream software, LSI Logic will enable system users to streamline business processes through inherent, easily customizable work flows that automate a variety of routine tasks. Work flows use messaging, next steps, and conditional and rules-based routing, as well as transaction delivery, to tailor the work-flow path within a department.

LSI Logic hired the Hunter Group, the consulting firm that specializes in human resources technology and processes, to work with it through implementation. For the early-release site testing, LSI Logic employed a version of the SmartStream Step methodology for planning purposes. SmartStream Step describes a set of activities and guidelines to follow and details how each should be executed. By providing standard deliverables and recommended techniques, the team was able to determine objectives, expectations, and a realistic time line for project completion. Key objectives were to install a client/server environment and gain technical and application expertise, then validate new processes with the work-flow feature of SmartStream.[19]

Expert Says the Internet Is the Most Important Development in Human Resources

The Internet is affecting human resources functions at corporations in many ways, but consultant David Link of the Hunter Group says it's actually having more impact than any other technological development. "The Internet reduces costs while it delivers better service to its customers." Noting that the Internet, intranets, and extranets will merge, he lists the many ways they're already being used:

➤ Listing job openings and providing information to applicants

➤ Accepting and qualifying résumés

➤ Processing new hires, including orientation

➤ Providing information on benefits and policies, including at external firms

➤ Allowing employees to change personal data

All these functions are time-consuming for human resources staff, and allowing them to be done automatically is eliminating rote work and drudgery. This can free up an enormous amount of time, allowing staff more time to think and interact with people, doing the higher job that is needed. The way in which HR staff treat this change will have a big impact on them—and their companies, he notes. HR is the last element of the corporation to leverage technology. The effect is to move HR from the back office to the front, where it can help drive company strategy. "HR is the last element of competitive advantage for corporations. With the shortage of qualified people in the market, how companies use their human resources will determine how well they succeed."

■ TECHNOLOGY IN HUMAN RESOURCES

Changes in technology used for human resources have been largely invisible to most employees up to now, but that may soon change. Payroll was one of the first applications companies automated, and now many other human resources issues are joining it. Staff costs are obviously a major expense at most companies, so it's not surprising that, as they move toward integrated enterprise management systems, payroll and human resources are often their first steps. Many companies are adopting these general-purpose packages, which often require extensive customization for a specific application. Others choose industry-specific packages, which are offered for both tiny and giant enterprises.

Those changes help the company, but the place employees will benefit is in using the Internet as an interface to these packages and other functions. Companies are rushing toward self-service schemes over the Internet and sourcing out benefit management to save money and provide better service to their employees. The Internet is also becoming a significant source of staff, particularly in technical and other specialized fields where the Internet is widely used. This—and conventional recruiting activities—are creating a glut of applicants, in fact, but again technology has an answer: automatic résumé scanning and screening.

Chapter 8

You Can't Keep Secrets from Finance

■ INTRODUCTION

Accounting and finance were the first applications for computers at most companies, and for many years computers were widely consigned to the accounting department. Though that may have slowed the use and integration of computers into other parts of companies, it meant that a vast number of people and companies focused on producing financial and accounting applications. This is true, from the smallest enterprise to the largest. Today's inexpensive PC packaged software such as Quick Books and Peachtree offer fully integrated capabilities even the biggest company couldn't have only a few years ago.

That was the era of the glass room, with big IBM mainframes dutifully accepting, digesting, and storing data, producing reports on seemingly endless sheets of Z-folded perforated paper. As companies demanded more and more from their computers, however, they had to keep buying faster and faster ones, yet they still couldn't keep ahead of the explosion in information. Instead, companies started turning to client/server technology.

The client/server revolution was fostered by powerful desktop personal computers, which individually couldn't match the power of the mainframe, but collectively far surpassed it in the hands of eager users. Soon clients—workstations on the desktop—were doing more and more of the work, with the mainframe relegated to a server that primarily stored information in a readily available form for the client computer to access. Other servers performed calculations or served other specific purposes. No longer were users captive to the people in the glass room, but they could use the computer to do what they wanted, whether it was to find information on trends or generate reports of customers. The process generally also means an immense change, because the financial and accounting information technology no longer stands alone. It's part of an overall corporate system.

The giant accounting software companies of the past, like Cullinane, have mostly disappeared or shrunk, victims of changes in technology. New names like PeopleSoft and Baan have replaced them. Some, like SAP, recognizing the trends, converted their business acumen and technology expertise to new client/server technology and have become even bigger in the process.

That is the big excitement in accounting in finance today. Established companies are replacing their obsolete mainframe applications with integrated, enterprise-wide applications based on client/server technology. For most companies, it has been a wrenching process, accomplished over a long time at great expense, with the assistance of outside consultants.

The changes have involved not only new hardware and software, but, usually, significant trauma for accounting and MIS staff used to the old ways of doing things. But like a root canal, the process was worth the pain. All agree that it was worthwhile when it was over, with lowered costs, better performance, and happier users of the information.

■ MIDSIZE COMPANY REPLACES MAINFRAME SOFTWARE WITH STANDARD PACKAGE

Collagen Corporation of Palo Alto, California, develops, manufactures, and markets biomedical devices for the treatment of defective, diseased, traumatized, and aging human tissue. The company has been in business for 20 years, has 400 employees worldwide, and generates $70 to $75 million in annual revenues.

Collagen wanted a user-friendly accounting system that ran on a client/server platform, integrated with other systems already in use, required little training time, and provided excellent reporting capabilities. Collagen chose Great Plains Dynamics C/S+, a financial package for Windows that offers an intuitive, integrated, and powerful process for record keeping, reporting, and distribution of financial information. The product provides import abilities and electronic-mail functions that reduce Collagen's need for paper, and its inherent flexibility and ease of use have improved reporting capabilities, increased employee productivity, and substantially improved service to the company's customers.

Collagen needed to replace its previous finance system, which had limited capabilities and was no longer supported by its manufacturer. The company also wanted to improve its reporting capabilities and provide managers with on-line access to accounts payable and internal entry information. Its goal was to implement a flexible, intuitive, and user-friendly system that would provide the company with faster and more detailed information, greater reporting capabilities, and lower maintenance costs.

The company runs the software on a Digital Equipment Alpha server under Microsoft Windows NT. User workstations run Microsoft Mail, Access, Word, Excel, PowerPoint, and Schedule+.

One attraction to Collagen was the products' drill-down capabilities, which allow users to track transaction trails throughout the system. Another was the intuitive Windows-based interface, which minimizes training time and increases employee productivity. The interface, plus modifiable work-flow checklists, lets users move through their tasks with a minimum of keystrokes, improving productivity and reducing errors. Ease of use has reduced both the need for information services support and the length of time previously spent in production and distribution of financial information. Import features allow Collagen to eliminate repetitive and time-consuming data entry. The drill-down feature allows managers and authorized staff to drill down through a report to the underlying financial data.

The company's old system, which was mainframe-based legacy software, no longer was supported by its manufacturer, and its reporting capabilities were almost nonexistent. Data entry, record keeping, and the production and distribution of reports were all completed manually, which was repetitive and time consuming. Generating monthly reports required the assistance of the information services department, and maintenance had become cost prohibitive.

To reduce maintenance costs and decrease the finance department's dependence on information services, Collagen wanted a software solution that ran on a client/server platform, with more ease of use and far greater reporting capabilities. Moreover, the system had to be easily modified from within the company, without the need of outside consultants. Collagen wanted a complete system that would provide quick access to financial information and would integrate easily with other systems used throughout the organization.

Installation of the software improved production and distribution dramatically. The finance department creates and customizes reports without the need for outside assistance. Recurring reports, such as monthly and quarterly statements, can be set up once and saved, providing for easy and consistent regular reporting. Single ad hoc reports, such as those needed for a specific project or situation, can be produced in a matter of minutes instead of the hours or even days they would take before. Reports can be sent via E-mail, thereby virtually eliminating the need for paper and reducing the time and resources spent to a fraction of their previous amounts. Perhaps most important is the ease and speed with which employees learn and are able to maneuver through the system.[1]

■ FEATHER COMPANY RAMPS UP TO REPLENISH CUSTOMER DEMAND

Pacific Coast Feather Company in Seattle is a $130-million consumer products company that supplies down pillows and comforters to virtually every major retailer in North America. It maintains a dizzying pace, even for the consumer products industry. It manufactures 2,500 unique products using 12,000 raw materials in a make-to-order environment with average turnaround times of three to five days, claims president Roy Clothier. "We have had to ship pillows to 40,000 different addresses within 24 hours of receiving the order, all while growing the business at a 20 percent annual rate and building a new plant a year for the past four years."

The bedding manufacturer picked SAP R/3 to keep up the pace. The company wasn't getting adequate information from its old system in a timely and effective way, which hampered its ability to stay on top of its business and maintain its service standards, states its MIS director Gwen Babcock. "With R/3, we have a single, integrated system that provides us with real-time information, so we can react more quickly and know where we're at on a day-to-day basis."

Being a fast-moving company, Pacific Coast implemented R/3 rapidly across its eight manufacturing sites in the United States and Canada. It went live in four months with financials, controlling, sales and distribution, and materials management. Top management support, a strong internal team, and supportive project consultants from SAP were the key to the speed of the implementation. With R/3, the company believes it is poised to handle its explosive growth, including doubling over the next five years without impacting service to its customers.

In fact, R/3 will improve that service. "We've already reduced order turnaround times by about 40 percent since we began using R/3," says Babcock. In addition to turnaround time, quality, marketing, and innovation are important to the firm's customers. It sells to retailers that place high emphasis on quality, but also expect it to be cost competitive; therefore, its manufacturing logistics have to be efficient so it can provide extraordinary service to customers. The company ships huge volumes of relatively small orders every week, replenishing retail store shelves on a real-time basis.

The situation is complicated by lead times for raw materials, which can stretch as long as nine months, so it relies very heavily on forecasts. Then it has to match incoming and fluctuating order rates against projections and adjust raw material buys very quickly.

The company's previous information strategy wasn't making the grade. It was based on individual systems at each site doing batch transfers of files on a daily basis. Corporate information was always a day behind and out of sync with the data at the individual plants. It needed a single, integrated system to provide better visibility into raw material inventory and procurement so it could react quickly when orders didn't match the forecast.

The company's search for integrated business applications software didn't take long. "The selection process was really made very simple for us, once we got a closer look at R/3," Babcock notes. "In our view, R/3 was clearly superior to any of the competition in terms of functionality, flexibility, and technical environment." She was particularly impressed with the depth and flexibility of the R/3 system. "R/3 is table driven—there is a great variety of parameters you can set to define your business processes. That depth of functionality allows us to define a business process for every part of our business. R/3 then provides us with the flexibility to change the way we handle the process by simply flipping some switches."

R/3's manufacturing prowess was also a big plus. It needed a system that could accommodate multiple factories and manufacturing environments. Its ability to work in more than one currency was also a factor, since the firm has a manufacturing plant in Canada and does business there.

More than 90 percent of Pacific Coast orders are computer-generated by customers and sent electronically via EDI. Its customers keep track of every item on their shelves, then apply an inventory replenishment algorithm to order from the company. The information moves directly into R/3 as a sales order and flows through production planning to schedule manufacturing on a make-to-order basis. Production planners also use R/3 to check available raw material inventory and release documentation for the shop. Pacific Coast then produces the product, "receives" it as finished goods, creates a delivery document, and prints the necessary packing slips and shipping labels.

Pacific Coast also transmits an advanced shipping notice and an electronic invoice to its customer through EDI. All that information then flows through to profitability analysis and the general ledger and updates the sales information system for that order. An enormous amount of information gets automatically updated, allowing the company to look at any order from many perspectives.

The software also allows the company to get a clear picture of profitability, what it costs to produce a product versus its standard costs, to determine whether it has met its margin goals. It can also analyze substitutions, examining the cost of substituting a higher-priced raw material—if the company is short on the one specified in the order—to meet a delivery date. Inventory is also more accurate. The company used to shut down its plants for three days four times a year to take a physical inventory, but the firm has been able to reduce that to twice a year.

Financials are also done more quickly and more accurately. Prior to R/3, it took three weeks to produce internal financial statements; today it takes four days. The firm is already reducing inventory at the same time that it is improving its ability to service customers.

The company plans to use R/3 to get a better understanding of customer-by-customer profitability. It provides customers with a lot of services beyond basic product, such as sales training, point-of-sale support, materials, packaging, sales brochures, and signage. R/3 allows the company to place a value on these services so it can justify its pricing.[2]

■ SAP COVERS TILE COMPANY'S NEEDS

Another company that switched to SAP is Monier, the leading manufacturer of concrete roof tiles in the United States. It is a subsidiary of Redland, an international company headquartered in England and the world's largest manufacturer of concrete and clay tile. Operating independently, Monier is based in Irvine, California. It has more than 800 employees and annual revenues of more than $100 million.

The company believes that its ability to respond to customer needs is the key to remaining at the top of such a highly competitive industry. This requires a flexible and powerful system capable of handling all the data involved in planning, controlling, and coordinating the processes that take place between procurement, sales, and delivery. The legacy system at Monier couldn't maintain the company's customer service. Access to financial information was difficult and time consuming, and inventory was out of control. Additionally, neither the legacy system's hardware nor its software was supported any longer, putting the company at great risk of losing all data should a breakdown occur.

The old system consisted of an IBM 9335 mainframe, Xerox XBMS database, and software written in the 1960s and 1970s. Historical data was simply not available. Financial, sales, and distribution information could take from 30 to 60 days to produce, which meant that the chief executive officer and his managers were not getting the information they needed to administer the business effectively.

After a comparison of 25 different software packages, Monier found that SAP's R/3 client/server system running on a Compaq ProLiant 4500 using Windows NT Server best met the company's needs, offering financial, asset management, and sales and distribution functions in one integrated package that required very little modification.

With 15 offices and manufacturing plants throughout the United States, Monier was looking for an operating system that could be managed from the central office, but would still afford each site a certain amount of autonomy. The system also had to be easy to learn and use, without the need for technical support staff at each site. Finally, the company wanted a system that would allow it to eventually move R/3 to a three-tiered topology.

Monier assessed its need for a new system with an eye to the future. It held a series of meetings with employees who would use the system. They assessed and weighed the problems within the organization. Employees felt that to be successful over the long term, they needed a system that would help them to develop a better view of market and customer activity, focus on customer responsiveness, and control costs of inventory and operations.

The system they got offered improved communication between financial and manufacturing processes; greater control over sales, distribution, and customer communication; faster and more detailed access to information throughout the company; and better tracking and management of inventory, leading to a 30 percent improvement in product delivery.

Using the system, managers can identify their top ten customers and the top ten products those customers purchase, allowing the company to evaluate its position in the marketplace daily. Bringing inventory under control has increased Monier's ability to deliver product on a given day, thereby improving customer service and significantly increasing sales.

The implementation was fairly smooth. Within six months after the selection, Monier implemented both the Materials Management and Financial modules. By the ninth month, the company implemented the Sales and Distribution module at selected locations.

Currently, 20 users in Monier's headquarters use SAP modules for finance, materials management, purchasing, and accounts payable. Another 140 users are on-line at the company's plant locations. The client/server platform consists of an office server with Windows NT Server and an R/3 database server, application server, and development server, all with Windows NT Server, and 15 network servers at branch locations on Compaq ProLiant 2000 with 64 megabytes of memory. The servers are connected via Cisco routers.

Monier's plans include bringing its human resources and payroll process on-line. The company also plans to migrate from the

Oracle database running on Windows NT Server to Microsoft SQL Server for Windows NT. Says Ed Bye, Monier's director of information technology, "We originally took the conservative approach and went with Oracle because, at the time we designed our system, there was not a large enough SQL Server installed base for us to predict installation and operation issues. Since then, I've discovered that SQL Server will give me better throughput—especially as my customer base grows."

As the new system rolls out to all Monier branch offices and manufacturing plants, Bye will evaluate the performance of the current database and then draft plans for the upgrade to SQL Server with the advantage of working with a single vendor. Long-term goals include integrating plant maintenance into the system.[3]

➤ The World's Largest SAP Implementation on Windows NT and SQL Server

What happens when two international software giants team up with a big six consulting firm to execute one of the largest-scale client/server migrations in history? When the players are Microsoft, SAP, and Deloitte & Touche Consulting Group/ICS (D&TCG/ICS), the answer is: a breakthrough legacy systems overhaul that's on time and on budget, and establishes a new standard for excellence in teamwork, project management, and technical integration.

By early 1995, the legacy systems supporting Microsoft's finance and administration had finally buckled under the pressure of the company's explosive growth. As Microsoft bolted on more client/server add-ons, the legacy systems became increasingly unwieldy, expensive, inflexible, and unable to deliver. Microsoft's challenge was to develop a unified general ledger, streamline and standardize business processes worldwide, establish worldwide visibility and control over capital assets, establish worldwide business performance standards, and eliminate the multitude of legacy systems and interfaces. Concomitantly, it overhauled its processes and implemented new systems for finance, management reporting, legal consolidation, procurement, accounts payable, assets, services billing, tax reporting, and integration with more than 30 external systems.

The SAP modules it installed were financial accounting (FI), controlling (CO), fixed assets management (AM), materials management (MM), and parts of sales and distribution (SD). Not surprisingly, Microsoft chose Windows NT and its SQL Server. The computers used are Compaq ProLiant 4500 computers, each with four 166-MHz Pentium Pro processors. The database server has 1 gigabyte of random access memory with a 144-gigabyte hard drive and a duplicate for backup. The application servers are seven similar Compaq ProLiant computers with 512 megabytes of RAM and 16-gigabyte hard drives. Clients use Windows 95 or Windows NT on 486 or Pentium PCs.

The project required quite a few people to complete: 19 from Microsoft finance, 22 from its information technology group, and 18 consultants from D&TCG/ICS.

The phased approach started in August 1995, with the first part completed in seven months, general ledger (GL) installation and Canadian implementation completed three months later, and 24 subsidiary implementations installed over the next ten months. The project resulted in:

➤ Improved management decision making

➤ Rich functionality and global scalability

➤ Client/server architecture on Windows NT and SQL Server with Microsoft Office integration

➤ Increased user productivity

➤ Dramatic cost savings from hardware and software consolidation

The decision to implement SAP R/3 was not taken lightly. Twice prior, first in 1992 and later in 1993, Microsoft evaluated a number of software packages including SAP R/3. Microsoft selected SAP R/3 because of its rich functionality and global scalability. Microsoft needed a sophisticated product that could handle a complex multinational structure and was committed to implementing on Windows NT and SQL Server to take advantage of the open client/server architecture, homogeneous operating system, and integration within the Microsoft BackOffice family, as well as with Microsoft Office products such as Microsoft Excel and Word.

Microsoft committed both business and information technology resources, providing a balance of business needs with technical support knowledge. This strengthened executive involvement and support, and let the team make key business process changes quickly—critical to the project's success.

Microsoft kicked off phase one in September 1995 to implement SAP R/3's procurement, accounts payable, asset management, and finance/controlling master data modules for 100 users at its corporate headquarters. This took seven months.

Phase two's focus was bringing up SAP R/3's general ledger modules on July 1 for the new fiscal year. This brought the total number of users to 1,300 and completed the transition of the financial structure to SAP R/3. Phase three was the worldwide rollout for 25 additional subsidiaries, making Microsoft the largest Windows NT and SQL Server SAP R/3 implementation in the world.

Microsoft established clear goals for its business transformation process: global financial standardization and streamlining of procurement processes. To reach these goals, Microsoft took advantage of the power of SAP R/3 software. According to Gregg Harmon, Microsoft project manager, "We didn't do a 'blue sky' redesign. We worked with the firm's Business Process Reengineering (BPR) services from the start, achieving process integration by following SAP's natural flow. We allowed SAP to drive the business transformation instead of the other way around."

The implementation provided an excellent showcase for Microsoft's application development technology. Microsoft's Object Linking and Embedding (OLE) automation is used to update SAP from Microsoft Excel using Visual Basic for Applications. It was relatively straightforward to create sophisticated data validation and upload tools with existing desktop tools and SAP's integration facilities.

For example, Microsoft works with several strategic vendors who are not EDI-capable. These vendors E-mail their invoices to Microsoft Accounts Payable as formatted Microsoft Excel spreadsheets including Visual Basic for Applications macros, which validate and upload the data to SAP, eliminating rekeying of data.

Microsoft employs the same mechanism to interface financial transactions from other internal systems. It still needs to post journal entries from about 50 ancillary internal processes and systems that have not converted to SAP. Over 70 percent of these transactions are entered into SAP through Microsoft Excel interfaces, increasing data accuracy and efficiency.

Perhaps the most compelling of these solutions is Microsoft Market, an Internet Explorer– and SQL Server–based application that allows Microsoft's 20,000 employees to submit purchase requisitions through the corporate intranet. Purchases from strategic vendors are made simple through the "shopping basket" metaphor. Links to SAP R/3 ensure that the user codes the requisition correctly. Finally, the application uses work-flow logic to route a Microsoft Exchange E-mail message to the appropriate approval authority for tacit approval before transmitting the requisition directly to the vendor. Only requisitions that require procurement intervention are routed into SAP. Besides transaction accuracy and speed, an added advantage over paper-based requisitions is the users' ability to inquire on the status of their requisitions through the Microsoft Market tool.

Microsoft plans to track return on its investment in the new software over time, but it is too soon to confirm actual figures. At a minimum, however, Microsoft expects a vastly improved management decision-making capability; significant hardware, software, and maintenance savings in its legacy system; and optimization of vendor price discounts and accounts payable discounts.

With SAP R/3, end users are already reporting increased productivity and the ability to handle more complex transactions. A common transaction such as creating a purchase order is much simpler today. With the old system, users cycled through half as many screens as SAP R/3. However, since the new graphical interface provides default data, list boxes, and prompts, the transaction is completed much more quickly.

Microsoft looked to SAP as a tool for streamlining and standardizing its financial business processes worldwide. A multitude of disparate legacy systems and nonparallel processes prevented the company from achieving its worldwide business objectives. For Microsoft, SAP has helped establish a single point from which organizational taxonomies, data hierarchy, and reporting can be managed. Teamwork between Microsoft, SAP, and D&TCG/ICS was the best way to accomplish those objectives rapidly.

While SAP R/3 helped establish more financial standardization and control at Microsoft, it exceeded expectations in its ability to support Microsoft's front- and back-end applications. Although Microsoft started with the finance and procurement implementation, its goal is broader benefits from implementing SAP companywide. The success with technology integration, coupled with the goals met on the financial side, makes it clear that other segments of Microsoft's business can benefit from SAP R/3. To that end, Microsoft is aggressively evaluating opportunities to use SAP for other processes, including worldwide operations and human resources.[4]

➤ Phillips Petroleum Uses SAP for Worldwide Business Platform

Phillips Petroleum also chose SAP R/3 in the largest IT project ever undertaken by the firm. Phillips, one of the world's leading integrated petroleum companies, first went live with corporate financials, procurement, and all R/3 logistics applications at Phillips Driscopipe, which makes high-density polyethylene pipe.

R/3 replaces a collection of IT programs on multiple platforms. "The implementation of the R/3 product suite throughout Phillips is designed to provide significant payback to the company in lowering our operations costs and providing better access to information for Phillips employees worldwide," said John Mihm, senior vice president of technology at Phillips.

The project encompasses implementation of all R/3 modules including financial accounting, controlling, fixed-assets management, project systems, sales and distribution, materials management, production planning, plant maintenance, human resources, and a specialized industry-oriented package called IS-Oil. Final configuration will include operations in the United States, Norway, Puerto Rico, Canada, the United Kingdom, and Singapore. The project will be delivered in five phases, with a final live date of July 1, 1999.[5]

■ BLUE CROSS/BLUE SHIELD UPDATES FINANCIALS

Of course, SAP isn't the only supplier of integrated enterprise financials. Blue Cross and Blue Shield of Louisiana chose Smart-Stream Financials from Geac, which bought the software operation from Dun & Bradstreet.

Just as sophisticated information systems are allowing doctors to better track patient care, especially in cancer treatment, Blue Cross and Blue Shield of Louisiana is applying new information technologies to its financial systems to support that care.

Blue Cross and Blue Shield of Louisiana, which employs more than 840 people statewide, recently began a transformation and a new direction in an industry dominated by mainframe computers. The company believed its systems improvement would cut the costs of financial administration, while meeting growth challenges faced by the organization. Its 100-day implementation of client/server and SmartStream financials has produced significant financial savings via staff realignment, cut invoice processing time by more than 30 percent, eliminated manual accounts payable record keeping, and stopped reliance on MIS for routine reporting.

Every cost factor in health care is coming under closer scrutiny today, and the cost of administering health care insurance is no exception. One of the most recognized icons in America today, Blue Cross and Blue Shield has for 70 years maintained its position as the largest health care system in the nation. Collectively providing coverage to one in four U.S. residents, the organization has led its industry with innovations designed to deliver cost-effective health care choices.

As is typical in the insurance business, Blue Cross and Blue Shield of Louisiana has traditionally used a mainframe to house 400,000 associated member files. The resulting volume of information can present an overload of paperwork.

When information technology began its transition to client/server as the preferred technology platform, Blue Cross and Blue Shield of Louisiana embarked upon a systems and technical review of its benefits to the organization. It soon charged Rusty Welch, its new director of budgeting and accounts payable, to streamline the accounting department. He was told to carry out the task "as soon as possible."

The first phase was to implement SmartStream Ledger, SmartStream Payables, and SmartStream Decision Support applications. "The concept and overall design of the SmartStream system is very effective for a health-care financing corporation," says Welch.

When Welch first walked in the door, most desktops at Blue Cross and Blue Shield of Louisiana held "dumb" terminals connected to the mainframe, and many employees were unfamiliar with personal computers. He soon brought in PCs and taught people how to use Microsoft Excel.

The overriding objective of Welch's work was the elimination of manual processes. In an office that handles 100 invoices per day, turnaround time needed to be reduced. Invoice processing took two to three weeks. It now takes five days, and the goal is two days.

In the accounts payable department, check writing was on a dedicated system. You entered the information, then the system wrote the checks. The data had no place to go other than the check, so it often "just went away," according to Welch. Limited to 30 batches of checks per month, the daily overflow was processed on typewriters. Then those checks were entered into the mainframe system at the end of the week. The work was documented manually, on paper and in folders—an antiquated, repetitive situation at best. The file room was bulging and the foot traffic was growing.

Other processes needed to be addressed, too, such as closing monthly journal entries. Nearly 200 entries were manually entered into a ledger book by employees. The book was then given to another person for entry into the mainframe. Yet everyone had access to the mainframe and could have done their work electronically the first time.

Another improved process was the preparation of fixed-asset journal entries on a spreadsheet. Without a great volume of assets, using a spreadsheet was not a problem. Yet the journal entries were still written in the ledger book by hand.

Just tracking an item description on an invoice required chasing through piles of paper. Because of the lag time in processing an invoice, and because an invoice would go through the hands of eight people, there could be three or four batches of invoices "floating around," according to Welch. They would have several hardcopy reports. They would look at those reports for the check number or the account number and review the account in the filing room or in somebody's office, then pull up the invoice.

Welch didn't have to look beyond his own national organization for a solution. He found satisfied SmartStream users at many other Blue Cross and Blue Shield operations.

Besides being confident that this was the right product for Blue Cross and Blue Shield of Louisiana, Welch's self-imposed time frame demanded software that was flexible and adaptable enough to be up and running quickly. Nothing would fit that bill better than SmartStream with its open architecture. An enterprise-wide business information solution, SmartStream is sculpted around open architecture designed to work the way businesses work. Work-flow technology facilitates the flow of data relevant to the task among users.

SmartStream Financials runs on a Compaq ProLiant 5000 server with dual processors, driven through the Windows NT operating system and Sybase SQL Server database software. Users are deployed with Pentium 75- to 166-MHz PCs with 16 or 32 megabytes of RAM. Consultants from Arthur Andersen helped expedite the project.

"We have found significant ease of use and simplicity using SmartStream on the Windows NT operating system," reports Welch. "In fact, we installed SmartStream in just 20 minutes on Windows NT!" The processing time is also exceptionally good with SmartStream and Windows NT, according to Welch. He uses his balance sheet in SmartStream Management Reporter as a prime example, producing in 6 minutes what used to take more than 30 minutes.

SmartStream was live and in production in 100 days, a fast implementation by any standards. "It takes most companies 13 months to do this," says Welch.

Blue Cross and Blue Shield of Louisiana began to see the outcome of the implementation unfold in short order. "The change to client/server has had a night-and-day effect," Welch says.

In payables alone, Welch's department gained the equivalent in hours of one-and-a-half full-time employees. Payables vendors number around 6,000, rendering around 2,000 invoices per month, with as many as 2,500 checks being processed—transactions that carry and generate reams of critical information.

The company's comprehensive enterprise reporting and analysis system, SmartStream Decision Support, allows users from all areas to access the information they need without going through an intermediary for reports. The applications share common features, such as business views for logically viewing and capturing only the data relevant to the task and a Microsoft Windows interface, with toolbars and consistent on-line help, making them easy to learn and use.

For example, quick ad hoc reports available using the Smart-Stream Query application are being produced from enterprise database information to answer day-to-day accounting questions. Those long treks through the file room tracking an invoice now stop at the desktop.

The system is configured so Blue Cross and Blue Shield of Louisiana users can go directly to the tables for information or drill down through the account balance window all the way to payables and eventually to the invoice. Welch is now in the process of planning an imaging system to complement SmartStream, furthering the move to stop paper flow at the desktop. The invoice processing time is down to the goal of five days, with one employee handling some 2,000 invoices per month.

SmartStream Management Reporter provides the ability to build, for example, underwriting gain or loss reports. This document shows all the different product lines based on income and claims activity, including the claims ratio. Previously an executive had to go to MIS for the reports, then enter the data into a spreadsheet, formatting it to the particular reporting need. "Now we build it the way he wants it," says Welch. "We're able to modify what we were doing on the mainframe. There's a lot of flexibility in changing a report on the fly. Executives go in, push a few buttons, move lines around, do their own type of analysis. It's a good analytical tool. It's a better report now."

Along with SmartStream Analyzer, in which structures are tailored to the organization, reporting users are able to examine the multidimensional relationships of data to find the information they need to do their work more effectively. Welch says he can write reports from Management Reporter, Analyzer, and Query. "On the mainframe you can't do that unless you know how to program."

With budgeting as one of Welch's primary responsibilities, new ways of creating budget variance reports have been implemented. Executive team members can now get more information, in the detail they request, and they can get it delivered directly to their desktops or laptops. Before SmartStream and Sybase relational databases, the variance reports were done from a database that tracked information for the previous three months. Now Welch has configured Analyzer and the database to deliver a year's worth of data to the process. Not only has the previous database system been eliminated, but so have three days' worth of monthly work.

Much remains to be built into Blue Cross and Blue Shield of Louisiana's new client/server environment. Beefing up the technology and improving processes for even less reliance on the mainframe and manual tasks is underway. The plan is to train the 20 users and increase their productivity while slowing down any net-new hiring.

Phase two of the implementation has already begun. This phase includes the installation of SmartStream Asset Management, SmartStream Allocations, and SmartStream Purchasing for the company's more intricate reporting requirements.

More cost savings are anticipated as further inroads are made toward a paperless office. Welch plans to go paperless on requisitions and reduce check printing through electronic funds transfers with SmartStream. "Along with the imaging system, you're talking about removing the majority of filing," he says.

One of the next steps is to deploy the six district offices around the state through a new wide area network infrastructure, while implementing SmartStream for the company's subsidiaries and bringing them on-line.[6]

■ CORPORATE FINANCE AT PFIZER: STRIVING TO BE "THE BEST IN THE WORLD"

Changes in the political and regulatory arenas, as well as heightened competition, are forcing the health care industry to face enormous shifts in its marketplace. To proactively address these dramatic changes and improve its cost structure, Pfizer, a global research-based health care company with projected revenues of approximately $10 billion in 1995, launched a major reengineering effort three years ago across its financial function. The vision was to make Pfizer's Corporate Finance group the best in the world, providing superior transactional support, quick reporting, and a decision support system that would give effective information to guide Pfizer's businesses. And that required a major shift in the company's information strategy.

As Russell Jones, Pfizer's corporate director of application delivery, corporate information technology, explains, "We needed to change our perspective on providing financial and accounting information from one of simply making transactions and keeping the books, to being able to add value in the availability of true information about Pfizer's business performance." Sun Microsystems and its software partner, Computron, were able to provide the comprehensive business solution for that vision.

Pfizer's previous system, a mainframe using a mixture of custom and packaged applications, was inadequate in three key areas. First, it was very difficult and time consuming to retrieve information for ad hoc queries used for business decisions. Second, that information was not readily adaptable to new business processes. Finally, some of Pfizer's financial processes were not completely captured by the existing software, requiring a paper chase to resolve questions as simple as whether an invoice had been paid.

At the end of the evaluation, Pfizer selected the total business solution provided by the alliance of Sun and Computron. A key factor for selecting Computron client/server financial software from more than 25 offerings was its EPIC/Workflow architecture. This provided Pfizer with the ease and flexibility to customize the applications to precisely match its business processes, both today and in the future. Building on their ongoing collaboration, Sun and Computron jointly demonstrated a smooth, well-integrated system of software and hardware that met all of Pfizer's requirements.

Sun's SPARCserver systems running the scalable, multitasking, multithreaded Solaris operating environment provided a platform that aligned closely with Pfizer's worldwide strategic computing goals, and it is the foundation on which Pfizer intends to build its mission-critical financial applications across the globe. Pfizer initially implemented the solution in its New York City corporate offices with Computron's N-dimensions general ledger and accounts payable, purchasing, accounts receivable, and fixed assets, together with the EPIC/Workflow application support package.

The solution has improved Pfizer's financial-based business processes in all areas. Transaction activities, such as invoices serviced per hour, have increased. The status of departmental expenses, invoices, and other instruments is readily available in real time to all groups. Electronic forms are automatically directed through the correct channels for review, approval, or action. Use of paper forms has decreased. And company-wide financial consolidation reports for senior management are now compiled more quickly at the end of each accounting period, a key goal of the reengineering program.

Best of all are the completely new capabilities. Jones reports that access to information has been phenomenal for all groups, from general ledger accounting to accounts payable and divisional financial analysts. Now every group can easily retrieve any or all data related to the company's ongoing financial activities. Work is in progress to implement the system in Europe and elsewhere. When totally implemented in two years, it will support more than 10,000 users.[7]

■ TECHNOLOGY IN FINANCE

Finance, of course, was one of the earliest applications to be automated. The big change in recent years has been abandoning old mainframe-based legacy programs for integrated enterprise systems based on the client/server architecture. These programs from vendors such as SAP, Baan, and PeopleSoft are based on industry-standard relational database management software such as Oracle or Informix. Though many companies install only certain modules, particularly initially, the optimum use is to include all the parts that give a company complete control and knowledge about its operations.

Integrated enterprise applications are also now available for the smallest companies. Low-end software such as Quick Books, M.Y.O.B., and Peachtree Accounting have remarkable capability, with modules that can handle all accounting needs for companies as small as part-time businesses or as large as millions of dollars in sales.

Develop Your Organization

■ INTRODUCTION

In the last two decades, a fundamental shift occurred in the relationship between corporations and their employees. As the companies rethink who they are and how they operate, they've chosen to focus on their strengths and limit areas where they are weak. In the process, they've sold off, spun off, and shut down operations, eliminating many employees.

The largest numbers of staff laid off have been in lower positions like manufacturing. In these cases, companies are increasingly buying products instead of making them, and turning to subcontractors instead of manufacturing the products themselves. They're also subcontracting other types of services—from security to warehousing to information technology—all reducing head count to concentrate on their areas of strength. Many professional workers, such as writers, computer specialists, artists, and even lawyers who once found comfortable and secure positions at corporations, now find themselves to be individual contract workers or employed by service firms that rent their services to former direct employers.

Corporate restructuring has cut deeply into middle management as changes in the company's operations, made possible by new technology changed the manager-worker relationship. Despite the trauma restructuring has caused in countless boardrooms and among hundreds of thousands of employees, organizations often emerged stronger, with renewed vitality and direction.

Nothing has had a greater impact on improving the renewed corporation than improved communications and the streamlining it has enabled. It's directly driven the flattening of corporation management structures: There's no need for layers of managers when it's easy for the top to let the bottom know what's important, and there's no need for levels of gatekeepers when the new technician or receptionist can suggest improvements to anyone accessible by E-mail. Communications has made it possible for people to work in new ways, including from home, and it's also allowed engineers in India or Israel to contribute on equal terms with their colleagues close to headquarters. And communications has created unparalleled opportunities for cooperation among employees through E-mail, collaborative tools, teleconferencing, and intranets.

Suppliers, buyers, and their customers can now easily communicate, which has far-reaching implications for companies. This reduces reaction time and can eliminate major mishaps and disasters. It also means, however, that any person or organization that exists only to serve as an intermediary is in jeopardy, as is anyone whose function is simply to collect or disseminate information without adding any value.

Corporations are adopting many new approaches to improving their organizations, but virtually all have two parts: a new look at their processes and new technology. Either alone can make changes; together, they revolutionize the way the companies operate.

■ INTERNAL COMMUNICATIONS AND THE NEW WORK ENVIRONMENT

One example of the way a major company has approached re-creating its work environment is the WorkPlace Advantage. Owens Corning's project to combine people and technology to change the way its people work, as well as the way they work together. It uses new technology to help increase productivity by fostering collaboration and the transfer of ideas, experiences, and knowledge through the availability of information.

The WorkPlace Advantage project focuses on key characteristics of the new work environment:

➤ *Mobile/global:*   This means the ability to work as a seamless global enterprise, even from mobile and remote locations.

➤ *Paper-free:*   The aim is to reduce the use of paper for information storage and transmission. For example, the new headquarters was designed with limited file storage space and few printers and copy machines. Electronic scanners attached to the network throughout the building aid in reducing the use of paper.

➤ *Teaming:*   This facilitates the ability to improve productivity through collaboration. The new headquarters contains dedicated teaming areas, many with built-in teleconference and videoconference facilities to support collaboration. Teaming rooms also contain PC projectors and multiple network connections. Over 100 miles of desktop fiber-optic cabling and 300 miles of twisted-pair copper cabling within the headquarters alone provide network access. Lotus Notes and custom workgroup software applications further promote teamwork.

➤ *Learning:*   Interactive kiosks, CD-ROM multimedia capability, an intranet, and a dedicated discovery center help promote organizational learning. Video, audio, and computer-based training help promote self-directed learning efforts. Sharing information globally increases the spread of best practices and allows the organization to learn.

➤ *Customer focus:*   Owens Corning wants customers to recognize its commitment to understand and anticipate customer's needs and exceed their requirements. Electronic commerce (EDI/EFT), Internet E-mail, and inquiry systems keep the company close to customers. Additionally, users can send and receive faxes directly from their keyboards and can directly access customers' Internet sites.

The technology employed updates and replaces the existing personal computer, data network, and communication infrastructure within Owens Corning by focusing on standard commercial technologies to enhance the work environment.

The WorkPlace Advantage information technology impacts 7,500 personal computer workstations at over 120 sites worldwide.

It was implemented within one year. The desktop and portable computers utilize standard software including Microsoft's Office 95 application suite. The global network includes seamless connection for users and corporate intranet, electronic mail, fax, and information exchange capabilities.

These key characteristics contribute to efficient and timely sharing of information, experience, and knowledge. Other related benefits include the capability to continuously improve various business processes and share these improvements throughout the organization. WorkPlace Advantage has already contributed significantly to Owens Corning's initiative to develop and implement common, simple business processes and systems. Project development teams communicate through E-mail, distribute presentation materials via E-mail prior to meetings, videoconference with remote participants, and take electronic minutes that are universally available via the network instantaneously.

The availability and cost of standard computing and networking technologies, such as Pentium notebook computers, multimedia capability, Windows 95 and "plug-and-play," are at the heart of the WorkPlace Advantage project. Using standard components reduces the risk of technology obsolescence. This effort would not have been possible two to three years ago, since the technologies were too unstable and expensive to implement on such a large scale.

One of the most immediate and visible benefits of WorkPlace Advantage has been the ability to implement other organizational, process, and system enhancements. Based on the WorkPlace Advantage model, Owens Corning was able to design, configure, and implement a Global Enterprise Application System from SAP in two years. The first phase was completed in less than ten months and a more comprehensive implementation six months later. This aggressive schedule would not have been possible without the communications environment created by the WorkPlace Advantage.[1]

➤ When You Care Enough to Install the Very Best

The millions of people who celebrate with Hallmark products have made the company's name nearly synonymous with greeting cards, but the company has thousands of other offerings: Crayola crayons, Magic Markers, Silly Putty, calendars, computer software, Christmas ornaments, television movies, and more.

Hallmark is an 87-year-old company that has become the number-one name in its industry. It has the world's largest creative staff, consisting of 700 artists, designers, writers, editors, and photographers. Its worldwide operations boast 19,600 employees. It publishes products in more than 30 languages and distributes them in more than 100 countries.

With such a diverse product line, it's important for Hallmark to project a common vision throughout its decentralized, worldwide organization. But until 1993, Hallmark had no enterprise network to promote the flow of information between departments. Some groups had their own local area networks, but nothing was connected. Employees couldn't send E-mail or share documents across divisional boundaries, and network support was handled within each department. Sometimes an employee in one department couldn't access a printer ten feet away because another department owned that printer. Everybody agreed that these barriers between groups were unnecessary and should be torn down.

In 1993, Hallmark received executive support to implement a true enterprise network. A three-month review of available solutions led to the selection of Novell's NetWare and Compaq hardware as the integrated network solution.

Hallmark now operates a mixed-platform environment of 6,000 client workstations, approximately two-thirds of which are Compaq Deskpros, the other third Macintosh. Its 60 NetWare servers are a combination of Compaq ProLiant, multiprocessor SystemPro, and ProSignia machines. The network also includes a number of Windows NT application servers, several UNIX servers running the company databases, and a number of mainframes.

Having an enterprise network brings all Hallmark's departments together. All the IT-related problems such as access to printers and servers have been resolved. And Hallmark's employees can collaborate on projects much more easily than before.

Hallmark also supports remote access using Novell Connect Services and AT&T's WorldNet Intranet Connect Service. This public network service gives employees working from home and salespeople on the road easy access to the company's information archives and E-mail system.

Hallmark plans to keep adding to the network's capabilities. Because NetWare is an open architecture, the company can easily add new services such as more powerful network management tools, better integration of legacy applications, and Internet/intranet capabilities.

Now employees around the world use Lotus cc:Mail over the NetWare network. Marketing and sales are better coordinating their efforts, creative services groups are sharing ideas more frequently, and the manufacturing and product groups are planning their operations more effectively. Productivity has grown, and Hallmark is better able to project a common vision across the globe.

In addition, its sales force benefits from better access to marketing and sales information when meeting with customers. Because salespeople are spread around the world, it was difficult in the past to keep them updated on the latest developments. Now the AT&T WorldNet Intranet Connect Service lets them connect with its corporate information database from anywhere, so they can receive immediate updates whenever the company announces new products and programs.

Another key benefit is that users now have a single source for technical support. Prior to the implementation of our enterprise network, each department had its own IT group to handle acquisition, maintenance, and support of hardware and software. Now the company has one single IT department to take on all responsibilities. This provides users with more efficient service because the department has a single knowledge base from which to draw for technical solutions. It can keep track of problems and make sure they don't happen again.

There was a unique and exciting element to the company's implementation: the dial-in service for salespeople was among the first of its kind using AT&T's WorldNet services. Several years ago, it equipped nearly 1,000 field sales representatives with IBM ThinkPad notebook computers and allowed them to connect to its corporate information databases to upload sales data, retrieve program updates, and exchange E-mail and other electronic communications. This effort revolutionized the way salespeople work; it has allowed them to spend more time with customers and less time straining to keep up with the latest developments at corporate headquarters.[2]

➤ Intranets

More and more companies are seeing a specific new communications technology as the ideal way to enable internal communications. The intranet is an internal "Internet" that connects its employees no matter where they are. Technically, it can actually be a subset of the global Internet, but protected from outsiders by technology and passwords.

Like many international corporations, 3M needs to distribute information efficiently to a global workforce—in its case, 70,000 employees in more than 60 countries. Over the past decade, 3M has used a variety of technologies for its electronic communications. In mid-1994, with the goal of simplifying information sharing and making its global workforce more cohesive, 3M began migrating to a Web-based intranet using Netscape software.

3M's mixed-platform intranet, which includes nearly 10,000 copies of Netscape Navigator, runs on Netscape SuiteSpot server software and uses Netscape Catalog Server to provide a cohesive view of 50 internal Web sites.

Using the company intranet, 3M employees can read company newsletters; view job postings, department sites, company information and news, stock prices, and organizational charts; search the company's on-line phone book; browse the home pages of special-interest groups; look up E-mail addresses in a relational database; and view a listing of divisional and business-unit servers.

3M chose SuiteSpot server software for its reliability and security, ease of maintenance and support (which reduce intranet-related costs), and superior functionality.

With years of experience sharing information via proprietary and text-only technologies like Gopher, 3M was quick to recognize the power of cross-platform intranet technologies. "Our goal is for the intranet to be every employee's first stop each morning," says Norm Hickel, manager of electronic marketing communications and Internet applications. "They'll find out important news immediately, no matter which platform they use, without the delays associated with paper-based communications." Without any internal promotion to date, the 3M intranet already receives 3,000 to 5,000 hits per day.

3M now posts the many newsletters produced by departments, divisions, and geographic business units on the intranet. Previously, with tens of thousands of employees, even the simple task of distributing electronic versions of newsletters was harrowing. After preparing newsletter content, publishers were spending a significant amount of time producing different versions for the various client platforms. Now the divisions are excited because they can publish a single document, put it in one location, and be done. The firm is cutting paper costs and saving labor costs associated with sending electronic documents to various platforms—and even saving trees.

3M initially chose Netscape Commerce Server for its intranet deployment because of its security features. Now the company is migrating to Netscape SuiteSpot because of the extended functionality it provides, as well as its enhancements over the previous generation of Web servers. About 50 different business units publish content using Hewlett-Packard 9000 series workstations and Netscape Enterprise Server, SuiteSpot's core Web server.

One of the intranet's more popular sites is a page maintained by an Internet special-interest group that publishes schedules of Internet and intranet classes and a listing of divisional and business-unit servers.

As 3M continues to extend its intranet, it has assembled a Webmasters' forum to brainstorm new applications. One idea being implemented is a Navigator-based fax-on-demand application that customer service representatives can use to send faxes to their customers. Another application will allow 3M employees to view and modify selected components of their human resources information.[3]

➤ Domino's Pizza Powers Intranet with Lotus Domino

Since its founding in 1960, Domino's Pizza has grown into a worldwide company with 18 distribution and supply centers and 1,200 franchise operators. Ensuring that every pizza delivered meets Domino's standards requires constant communication between the company and each of its 5,600 stores. At the same time, keeping its far-flung network of stores up to date on the company's latest product and marketing efforts, policy changes, and service enhancements represented a significant challenge, particularly given Domino's reliance on independent franchise operators to meet the commitment to customers.

To support this process, Domino's created a global intranet to communicate with its stores around the world. The intranet lets it keep store owners and managers better informed and also automates applications that allow Domino's to improve operations while cutting costs.

Domino's wanted a system that would be easy to set up and maintain, yet be adaptable so it could add functionality as 50 new applications came on-line. Domino's wanted a controlled implementation, to quickly achieve near-term benefits while laying the groundwork for a more comprehensive system in the future. Moreover, it wanted a system with sufficient security to allow it to control exactly who has access to what information.

The first step was to move its company phone directory on-line. As with many companies, Domino's directory is a huge document that requires constant revisions. An on-line version could be easily updated and immediately accessible to all employees. The company also planned to move its internal newsletter on-line.

However, it also had more sophisticated tasks in mind. For example, the company wanted to create a detailed, interactive product information database with the latest information on its products. This information was contained in a dedicated, dial-up bulletin board system consisting of a series of hypertext-linked documents.

While the initial focus was improving communications, longer-term goals include moving mission-critical systems on-line. The company plans to use its intranet to integrate and automate the collection and analysis of accounting information. Currently, individual store and franchise results are sent in by fax, telephone, and mail to the company's home office for consolidation and analysis. Consolidated summaries of this data are then sent in hard copy to each franchise. With an on-line system, franchise managers can easily upload their latest performance information to Domino's central server and quickly view how their stores rank against Domino's worldwide operations. To do this, Domino's needed a system with security features that would allow it to control exactly what information could be viewed.

And finally, Domino's Pizza wanted to provide on-line resources for employees and franchisees, including discussion forums and a document library, that would help to create a greater sense of community.

Finding a solution that would allow Domino's to fulfill all these ambitions meant finding a secure, flexible, and scalable Web application server capable of getting the company up and running quickly, while setting the stage for more ambitious projects to follow. Domino's Pizza initially looked at both Netscape and Microsoft servers to drive their corporate intranet. However, in both cases, the company found that these systems lacked the security, manageability, and legacy system support that Domino's needed. Moreover, with plans for over 2,000 Web pages, the company needed a system that was easy to maintain and update. In the end, Domino's Pizza turned to Lotus Domino.

Working with Fry Multimedia and MicroAge, Domino's developed the necessary databases and the Domino-enabled intranet to meet its initial objectives. The Domino's Pizza intranet is accessible to approximately 1,000 employees and is initially focused on the 18 distribution centers that provide ingredients and supplies to franchise operators.

Domino's Pizza has already developed six Lotus Domino applications to handle its specific information needs and plans to develop over 50 more. In addition to its phone directory and employee newsletter, the company has created applications that allow employees to check calendars, view policies, and download mission-critical information. The system contains a detailed document library, which gives employees easy access to a broad range of company information, such as corporate accounting procedures and calendars.

Using Lotus Domino's built-in interactive discussion capabilities, Domino's Pizza has created what has become its most popular application: an on-line discussion forum that covers topics ranging from product distribution issues to human resources management.

With this first success, Domino's Pizza is beginning the second phase of its intranet development, which will make even greater use of Lotus Domino's interactive capabilities. One specific application will automate an on-line financial reporting system, using Lotus Domino to link directly to the company's Informix data warehouse. Domino's Pizza envisions dozens of future phases, which will continue to add more applications to the intranet.

With its ability to easily handle massive databases of thousands of dynamic Web pages, Lotus Domino provides a simple means of managing even the most complex intranet site. Paul Messink, manager of desktop applications says, "Lotus Domino helps us achieve our goal of implementing our intranet as a manageable collection of databases rather than thousands of static HTML pages."

The company directory is one example. All managers and employees above that level have a personal home page with a photograph and relevant information. Instead of having to use HTML programmers to develop each of these pages, a Domino form automatically creates each page. When it wants to change a header on these home pages, it makes one small change to the Domino form and all the employee home pages are immediately updated. This ease of manageability also complements the company's goal to not have to train anyone in HTML. Domino's Pizza wants its information managers to manage information, not write code.

The Lotus Domino–driven intranet also offers the security that Domino's Pizza needs. With the goal of ultimately giving all corporate employees and franchisees access to the intranet, Domino's Pizza wanted a system that would allow it to easily control who had access to what information. For example, franchisees need to have full access to product inventories and shelf-life figures, but restricted access to pricing data. With Lotus Domino, the accounting database can hold all of this information, but permit access only to those pieces that the user is authorized to view.[4]

➤ Columbia Saves Millions per Year with Intranet

Columbia/HCA Healthcare is the world's largest health care provider, with 347 hospitals, 180 home health agencies, 125 outpatient surgery centers, and 75,000 affiliated physicians. It's always looking for more effective ways of conducting business and servicing its member hospitals, home health agencies, outpatient surgery centers, and affiliated physicians. To help cut costs and improve communication among employees and to customers, Columbia deployed Netscape client and server software for both a corporate intranet and an external Web site.

Columbia's intranet provides an up-to-date corporate directory, generates a variety of reports, posts physician résumés submitted via the Internet site, and helps train employees on processes.

One of the most popular medical Web sites, with more than 600,000 hits per month, Columbia's Internet site provides visitors with articles about health issues, downloadable healthy recipes, a directory of Columbia facilities, on-line magazines, and a schedule of physician chats that Columbia holds on an on-line service. Columbia also leverages its Internet site for various extranet applications, such as advertising surplus medical equipment to Columbia-affiliated facilities.

To run both its intranet and Internet sites, Columbia uses Netscape Navigator client software and Netscape Commerce Server, Proxy Server, and News Server software. On-line report distribution has significantly decreased costs, and though the figures aren't available, the potential for a $20 billion corporation is substantial.

Columbia's intranet site is served by dual Silicon Graphics servers. Employees access the site using Netscape Navigator over the existing LANs and WANs that connect all member hospitals. The Silicon Graphics servers came bundled with Netscape Communications Server. Columbia also purchased the Netscape Bundle, which includes News Server, Mail Server, and Proxy Server. The servers access additional database applications including SQL Server, Oracle, and FolioVIEWS on a Microsoft Windows NT server.

Columbia's intranet strategy leverages the popularity of its Netscape-based Internet site, which was named one of the top eight by *U.S. News & World Report* and received the highest rating to date from Six Senses, an on-line review of medical Web sites created by Echo Strategies Group, a medical communications consulting group.

Columbia leverages the popularity of its Internet site for various intranet applications, such as advertising surplus medical equipment. When one medical facility has surplus medical equipment, it posts a notice on the intranet. If no Columbia facility has requested the surplus equipment after 30 days, an automated process transfers the equipment listing to the company's Internet site, making it available for purchase by hospitals outside the Columbia network.

A second Internet and intranet application is used to recruit physicians. Practice opportunities are posted on the Internet site. Résumés submitted on-line are posted on the intranet. Hospitals can then approach the health care providers whose qualifications and location match their individual requirements.

Columbia established its Internet site in April 1995. Initially it considered the Internet site a form of advertising and used an Internet service provider to host it. Then the company expanded to using the Web to become a premier information provider. It provides the full spectrum of health care services, from information that consumers need to make informed decisions on health care to facilities that deliver services.

Columbia serves its Internet Web site on the same platform as its intranet site: dual Silicon Graphics servers running Netscape Communications Server, Commerce Server, Proxy Server, and News Server software. It uses a dedicated T-1 connection, and the server receives 600,000 hits per month. To ensure patient confidentiality, the Web server is separated from the corporate network and is protected by firewalls.

Columbia says it wants to stick to its core business: providing the highest-quality health care. It looks to partners such as Netscape to provide technology solutions. Its intranet is the core of a process by which its employees share expertise in health care with each other and its consumers. Ultimately, its goal is to help consumers make better decisions about their individual health care needs.[5]

➤ Teleconferencing

Today's intranet is based on text and still graphics, but it already has the ability to include audio and rudimentary video. It may soon evolve to a full-fledged teleconferencing network. That's something many people and corporations are anxious to have happen, because teleconferencing—both audio and video—has proven to be of enormous impact in improving corporate productivity. Figure 9-1 confirms the rapid growth in teleconferencing.

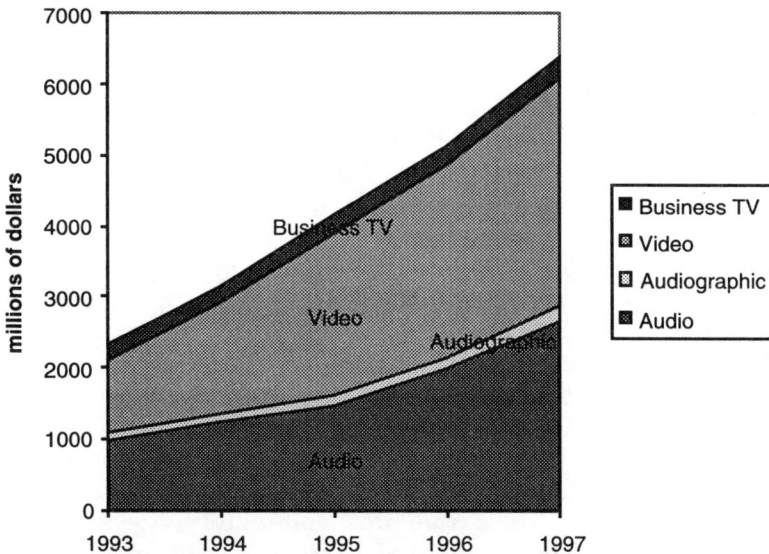

Figure 9-1. Teleconferencing revenue is growing rapidly, according to the International Teleconferencing Association.

It's a rare corporate employee who hasn't been involved in a teleconference. "Conference calls," as the telephone company calls them, have become big business. According to the International Teleconferencing Association (ITCA), an industry group, 5.1 million audio conference calls were held in 1996, for example.

These calls include everything from discussions with prospective customers to disclosure to stock analysts. They can be made from any phone—some by pushing a few buttons—but specialized service bureaus make the process even easier.

Audio conferencing seems to work best among people who've already met, but its biggest limitation seems to be the lack of graphics. Most meetings seem to require charts, tables, or pictures, and, though schemes exist to display information simultaneously on computer screens, most people seem to depend on faxed copies of such material. The Internet would be ideal for this, but the slow connections used by many people outside big companies still limit the quality available.

Most personal computers sold today have the basic capability for personal video, needing only a camera and software, so it won't be long before videoconferencing over the Internet becomes widespread, even if it involves only a small, jerky picture in a corner of a screen.

Among the providers of these systems is VDOnet, a new company in Cambridge, Massachusetts. Its VDOPhone offers videoconferencing via the Internet over regular phone lines or any TCP/IP connection—from dial-up modems through cable modems. Because of its dynamically scalable video technology, the quality of the video improves as the connection bandwidth becomes wider. Moreover, because the company has optimized this VDOPhone to take advantage of new technology, such as Intel's MMX multimedia-enhanced Pentium microprocessors, and Direct Draw support, its video quality and frame rates have been increased over what was possible even a short time ago. On intranets carried over high-speed local and wide area networks, it can provide full-screen video at up to 30 frames per second with echo-free speakerphone audio. It meets industry standards for videoconferencing, including interoperability between videophones from different manufacturers.

The ideal, of course, is true, broadcast-quality videoconferencing. That requires sophisticated and expensive equipment, such as satellite connections, but some companies, such as Tandem and Hewlett-Packard, have installed such systems. For most companies, a more practical alternative is dedicated corporate videoconferencing. The ITCA reports that about 150,000 hardware-based desktop video systems were sold in North America in 1996, compared to 12,000 group systems. The installed base of group systems is now more than 40,000 units. Each requires a special high-speed link, ideally T-1 at 1.5 MHz or at least ISDN at 56 kHz or 128 kHz, generally limiting use to larger firms and service providers.

■ TECHNOLOGY HELPS MANAGE THE MANAGERS

Though most applications of technology are to help manage processes, some actually are designed to help improve management skills. One example is software that actually helps managers define and communicate company goals, then set expectations and monitor the progress employees are making toward those goals. At Kayser-Roth, the supplier of No nonsense pantyhose and many other products, managers use Avantos ManagePro for this purpose.

The company CEO sets yearly goals, but it has not always been easy to tie that to individual work. The company first installed the product in its MIS department, where a director used it to focus on his group's contribution to better and faster customer service, one corporate goal. The first step was to define and analyze the existing systems and collaborate with management, which led to many system-related projects assigned to various project leaders and programmers. Defining the problems in this way helped provide focus, overcoming a common problem at many companies: constantly throwing new priorities at employees with no follow-up on previous efforts.

ManagePro is an integrated program that provides various views of the data needed, including a goal planner, a goal status board, a people status board, an action list, and progress notes. The top-level goal planner provides a hierarchical view of key objectives and subobjectives that support them. The displays of data are color-coded to highlight on-track items in green, those that have fallen behind in yellow, and critical situations in red. The firm uses a multiuser version of the product, which allows managers to keep individual data, yet share it selectively with others. It also is integrated with standard E-mail to allow employees to easily communicate project status. Avantos also offers an enterprise-wide version of the products. The program clearly requires a change in the way managers operate, and even if they agree in principle that it's an excellent approach, many tend to slip into old habits unless there's a strong commitment from above—software won't help managers manage if they don't use it. After the trial with MIS, Kayser-Roth extended use of the product to quality control and human resources, with plans to extend it to manufacturing and sales.[6]

At Claremont Technology Group, an unusual application is designed to help staff "manage" their supervisors. Using Lotus Notes on the company's intranet, the 600 employees can give feedback on leadership, management skills, and personality. As described in *Computerworld,* the system isn't directly tied to compensation, but to helping the managers improve.

To use the system, managers send lengthy questionnaires to employees, who fill it out and respond anonymously. The forms have to be sent to at least five employees to ensure confidentiality. Both staff and managers have found it a good way to improve the managers' skills, even to help the supervisors decide on future career paths. Some realize that they'd prefer technical work rather than management, for example, but in the end, that might improve management overall at the company.[7]

■ TECHNOLOGY HELPS IMPROVE PRODUCTS: PERFECTING ENGINES VIA VIDEO

Ranked 132nd on the Fortune 500 list of U.S. manufacturers, Cummins Engine is a leading manufacturer of diesel engines and related products. Its engines range up to 2,000 horsepower in size and power a wide variety of vehicles, vessels, and power generation equipment. The company has operations around the world, and to improve engineering and manufacturing productivity across time zones and to maintain a competitive edge, Cummins installed a VTEL videoconferencing system at facilities in Columbus, Indiana; Daventry, England; Huntsville, Alabama; Jamestown, New York; and Fridley, Minnesota.

With product engineers split between sites, Cummins Engine has used the video system for many purposes. It used it to help it comply with U.S. Environmental Protection Agency (EPA) emissions standards, for example. Engineers at plants in Jamestown and Daventry, where existing models are manufactured, consult regularly with their counterparts at headquarters in Columbus, where design engineers develop the next generation of engines.

The video systems are used for concurrent engineering between 80 and 90 percent of the time, and are especially valuable when staff need to view pictures or wear patterns on engine parts. As an example, on one occasion it allowed Columbus and Daventry engineers to resolve a critical production issue by examining a part, identifying the problem, and creating a solution within a matter of hours. The group saved at least 48 hours of potentially lost production time because it could meet spontaneously and share expertise—without requiring team members to fly from England to Columbus. This ability to maximize manufacturing time and minimize wear and tear on personnel is increasingly valuable to companies like Cummins with multinational operations.

After the success of Cummins' initial network, the Onan Corporation, a subsidiary that designs and manufactures generators for large-scale, specialized mobile trucks and recreational vehicles, added its own system. Its headquarters in Fridley uses videoconferencing for concurrent design with its manufacturing plant in Huntsville—and for meetings with corporate HQ in Columbus.[8]

Videoconferencing has helped Cummins reduce travel time (and related expenses) and, most important, has improved productivity in the competitive, time-critical industry of heavy engine and component design. Other corporations that few people would choose as technology leaders do the same.

➤ Railroads Depend on Videoconferencing

A giant in the transportation industries, Norfolk Southern Corporation is a $4.6 billion holding company that includes Norfolk Southern Railway Company and North American Van Lines. Its railway lines extend more than 14,700 miles through 20 states, carrying coal, automobiles, foods, and other products, and its motor carrier serves the entire United States and Canada.

To enhance relationships with its business partners and improve internal productivity, Norfolk Southern installed a VTEL videoconferencing system with document capabilities such as Pen Pal Graphics, fax, and slide management features in Atlanta, Georgia, and Roanoke and Norfolk, Virginia. It installed a multiple access control unit in Atlanta to handle multipoint conferencing plus receive-only stations at other locations. This allows viewers in these sites to see both other sites when holding three-way conferences.

Norfolk Southern uses the videoconferencing systems for internal departmental meetings, quality team meetings, training sessions, planning sessions, meetings with vendors and customers, marketing reviews, information sessions, and meetings with other railroad companies. The company uses the conferencing systems to enhance internal and external communications, and its conference rooms are consistently booked more than 30 hours a week. One corporate goal is to partner with customers and with other railroads and to accomplish the latter, it joined the American Association of Railroads' videoconferencing network.

Norfolk Southern bought videoconferencing systems expecting productivity gains, and it didn't take long to achieve them. On one occasion, several vice presidents based in Norfolk missed a scheduled company flight from Roanoke to Atlanta for an important meeting. The meeting was held by videoconference instead. On other occasions, marketing staff have been able to avoid the full day of travel necessary to reach two-hour meetings in Roanoke (caused by changes of flights in Charlotte or Baltimore) by using the systems. VTEL systems have helped speed the flow of important information at Norfolk Southern, which pays off in timely decisions and the ability to respond more quickly to customer needs.[9]

➤ Videoconferencing at Herman Miller

Herman Miller is a leading manufacturer of office furniture and systems. The company has been repeatedly cited as one of America's most admired corporations. Its products have been featured at leading design institutions, such as New York's Museum of Modern Art and Whitney Museum, and at the Smithsonian Institution in Washington. Herman Miller's beliefs in product design, people, and ecology have led it into more than 30 countries and to over $1 billion in annual sales.

Miller had a number of goals in mind when it established a video network. It wanted to bridge the gap between sales offices, regional operations, and key personnel located at its headquarters in Zeeland, Michigan. It wanted to provide a vehicle for communication between primary design and manufacturing sites located across Michigan and Illinois, and it intended to develop a network of public sites for meeting with suppliers and designers throughout the world.

A primary goal of the network is to extend the channels of participative management. Corporate personnel connect with major clients attending product presentations at Miller's Chicago sales office. Designers, service representatives, and sales representatives meet with clients to answer questions and obtain commitment for sales and joint ventures.

Herman Miller's staff comprises cross-functional work teams and department members who cooperate to provide their own expertise. One example is a "Chair Team" comprising members located in Zeeland, Grandville, Michigan, and Chicago. Its marketing team members exchanged strategies for catalog and brochure design. The company determined that the three Chicago and four Grandville team members saved over 300 hours of travel time to Zeeland during this three-month project.

The research and design department often meets with off-site suppliers for very technical meetings about materials and construction of furniture. At one videoconference held with a Minneapolis supplier, Herman Miller team members used flip-charts to display product schematics, the ergonomics of design. As the Herman Miller team outlined its research, it discovered that the information was too technical for the Minneapolis participants and a joint decision was made to redefine its role in the project. If this meeting had been conducted using an audioconference, the technical translation would have been lost, and the collaborative decision making might not have taken place.

The company conducts about 20 meetings monthly. Designers anticipate that videoconferencing will replace their frequent international travel. An ongoing function of the research and design department is meeting international product standards. The governing institutes are located in Nuremberg, Germany, and London. To be effective with new design reviews, Herman Miller conducts a series of meetings, and videoconferencing to these locations replaces costly weeklong international visits during the testing and certification phase.

Herman Miller work teams are realizing the benefits of increased communications between facilities in less travel time and travel cost savings. Members no longer spend time away from the office to meet with manufacturing and design team members. Videoconferencing meetings are also more efficient. When they're done, people go back to work. "Team members predict videoconference use will grow to encompass all aspects of Herman Miller's design and manufacturing business.

At Herman Miller, videoconferencing has provided an efficient tool for resolving time and distance barriers, and, more important, it allows project teams to reduce local and long-distance travel, thereby contributing to the environment-saving policies of the company.[10]

➤ Designing Products Long-Distance

Design Continuum is one of the world's largest independent design firms. It has nearly 100 employees located in Boston, San Francisco, Milan, and at an Australian affiliate, Adelaide-based Applied Design Development. The company's managers consider it an innovative firm with the entire world as its market.

Design Continuum was hired to develop a product and a user interface for three large, well-known Japanese companies. Its San Francisco office closed the contract by selling the clients on videoconferencing as a tool for collaboration. The design work was performed primarily at the Boston office with frequent video calls occurring over the course of the project. The project concluded with a final presentation meeting, which included a multipoint call to four sites—two in Japan plus San Francisco and Boston. The meeting helped to present Design Continuum as a unified company with coast-to-coast coverage in the United States.

Videoconferencing has had many beneficial effects on the company. It has expanded its markets far beyond geographic boundaries, allowing clients in Japan to feel comfortable working with a company on the other side of the world. It also demonstrates Design Continuum's prowess to leading-edge customers. High-tech Japanese companies were impressed by the technological feat during two-and-a-half-hour presentations.

Another way the company uses video is to find high-level employees. Since every employee meets customers, finding the right staff is a critical issue for the firm. Employee image helps a client to entrust a multi-hundred-thousand-dollar contract to the firm. So when the San Francisco office hires a new employee, Boston personnel participate in the interview process. But flying personnel from coast to coast to conduct interviews is expensive, so Design Continuum tried a virtual interview. In one such session, it was clear that a particular candidate was not dynamic enough, so he wasn't flown to Boston for the final round of interviews, saving the company the cost of travel.[11]

➤ Communicating beyond Post-its

One of the cleverest simple communications tools to appeal in decades was the Post-it note, the partly sticky notes that can be pasted anywhere as reminders. The inventor of the Post-it, 3M, uses more high-tech communications methods as well. The $15 billion company is headquartered in St. Paul, Minnesota, but the firm employs 85,000 in operations in nearly 60 countries around the world. 3M is famous for its research and innovation in developing new products, and videoconferencing helps maintain that reputation.

Using videoconferencing allowed 3M to create a virtual research lab. With video, one team goes to work and picks up where another team left off, even if they are thousands of miles apart. This kind of teamwork lets 3M develop ideas faster than ever before. Teams that were once limited to their own locations now operate across time and space.

At 3M, videoconferencing's value is most apparent in the final weeks before a product launch. By using multipoint videoconferencing, all teams are able to meet at the same time, which results in smoother, faster, and more efficient product launches.

One example is the way plants work together. Videoconferencing lets teams in two sites react and resolve issues swiftly when production problems come up. It also helps reduce product rollout cycles. With video, issues that once took nine or ten trips to resolve prior to a product launch are taken care of in considerably less time. For example, 3M's White City, Oregon, plant manufactures products for the company's Visual Systems Division in Austin, Texas. The White City plant must communicate to the division results on each early product batch to make sure that they meet final standards, so trial tests are conducted both at White City and in Austin. Teams at both sites meet regularly over video to discuss test results and to plan future trials. The manufacturing plant also relies on video whenever a problem arises during a production run. In certain cases, all production ceases until the problem is solved, necessitating an immediate solution. In these cases, video spares 3M lost time, money, and labor. In all, 3M estimates that video typically helps its White City plant cut downtime by 48 workdays per year.[12]

➤ An Aircraft Manufacturer Flies over Video

One company that could probably arrange meetings of far-flung staff, vendors, and customers more easily than most is Bombardier International, the diversified Montreal-based company that owns such aircraft makers as de Havilland in Toronto, Shorts in Ireland, Lear Jet in Wichita, Kansas, and Canadair also in Montreal. The company has production facilities in North America and Europe, and sells 90 percent of its production outside Canada. Yet Bombardier, too, uses videoconferences to reduce the time and expense of travel.

After a successful video trial, Bombardier's transport division bought five PictureTel System 4000 group systems for La Pocatière and St-Bruno in Quebec, Barre in Vermont, and Thunder Bay and Kingston in Ontario. These sites were connected using 112-kbps network services provided by Bell Canada and its partners. Bombardier's main application for its initial phase was to increase collaboration between managers in different locations. The video technology allows the company to save travel expenses and time spent when personnel travel. With videoconferencing, the staff members can also make quicker decisions, which in turn lowers production costs.

Bombardier's next foray into video allowed the company to perform data conferencing. With PictureTel technology combining audio, data, and video information, interactive document sharing has become an important part of its teleconferencing. After the videoconferencing proved its worth, Bombardier installed systems in its other major locations, both in North America and in Europe.[13]

■ IT'S VIRTUAL WORK: TELECOMMUTING

Of all the trends technology is fueling, few affect employees' lives more than working at home. While there have always been people who worked from their homes, changes in attitudes about work and technology are causing an explosion of home workers, including those who work directly for companies, those who are basically contract employees, and true independent workers.

According to a study by research firm FIND/SVP (Figure 9-2), more than 11 million Americans now telecommute to their offices. That's up 30 percent from two years ago, 175 percent from 1990. The average telecommuter is just over 40, earns $51,000 per year, and works about 19 hours per week from home. Of the 11 million telecommuters, 7.7 million were conventional employees and 3.4 million were contract workers.

Downsizing of corporations plus increasing preferences for lifestyle rather than job as the most important factor in a person's

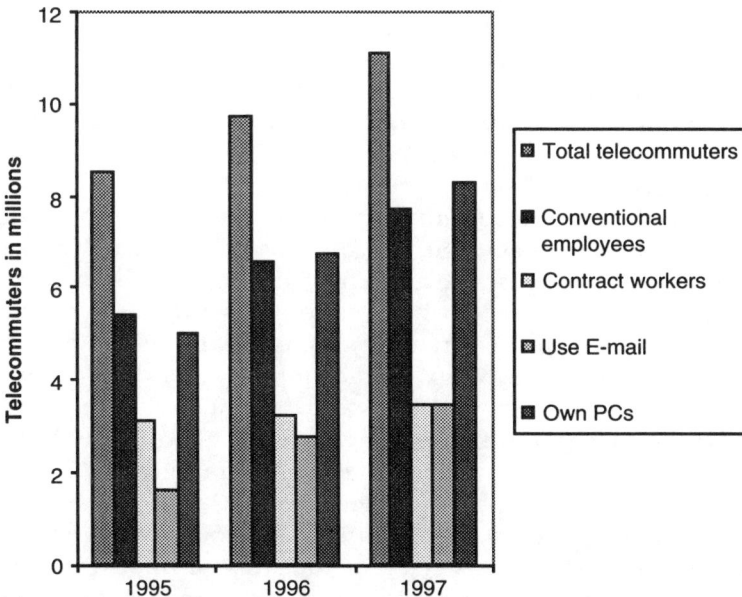

Figure 9-2. U.S. telecommuting trends from a survey by FIND/SVP.

life have been major factors in this movement, but it's technology that's making it practical.

Today, sophisticated telephone systems, inexpensive fax machines and copiers, and powerful but economical computer equipment and communications links make it practical to work away from a conventional office or factory, whether at home or even in a remote office. Of all these technologies, however, it's the Internet that will likely make the most difference. All three of its incarnations have impact—the openly accessible Internet, the company-focused intranet, and the wider community of customers, suppliers, and other partners collected into the extranet. They will tend to merge into one, separated by passwords and security, but however they're organized, they allow someone alone to function efficiently, even accessing information once impossible to find without immense effort.

The American Management Association found that 57 percent of companies use the Internet for electronic mail and 53 percent for information retrieval. And the FIND/SVP study found that 75 percent of telecommuters use personal computers, up from 59 percent in 1995. The research group also learned that 35 percent of the telecommuters use the Internet, including 31 percent who use it from their homes, for a total of 3.4 million users.

A study conducted across Canada by KPMG among 1,600 large and medium-size companies and 425 public-sector organizations reports that 4.5 percent always have telecommuters on staff, while 26 percent do occasionally. The study found that the leading reason for initiating telecommuting was to improve employee lifestyle, but once in place, the biggest advantage turned out to be greater effectiveness in use of time. The biggest disadvantage is lack of personal contact, which companies overcame primarily with occasional in-office meetings, plus extensive use of E-mail, teleconferencing, and other methods. Managers didn't find any particular management problems arising, and the companies provided their telecommuting employees with the same benefits and liability coverage. The majority of the telecommuters (77 percent) are full-time employees, 13 percent are part-time, and the others are independent contractors. Companies involved seem enthusiastic about the process; 14 percent are actively promoting it, and 20 percent have embarked on formal pilot programs rather than dealing with the issue on a case-by-case basis.

Today, in fact, customers and fellow employees hardly know where a person works. Unfortunately, if the technology is here, processes and society may not be. Many managers aren't comfortable unless they can see and interact with their employees. It's obviously not a universal solution, because many jobs require a greater amount of interaction or a physical presence, and some people need close supervision and the socialization of working with others, but telecommuting can be a very efficient system for both companies and employees. The companies can save substantially on facilities and equipment. Perhaps they can even institute standards for productivity and performance that can be measured objectively, though labor laws designed to protect the employee of yesterday from exploitation may prevent both employees and employers from benefiting from today's changed conditions. For employees, of course, benefits include eliminating the cost and time of commuting, savings in food and clothing costs, greater flexibility in work style, and greater focus.

Many companies have experimented with telecommuting, but many of the examples are on a relatively small scale. Not surprisingly, high-tech and communications companies have been pioneers in telecommuting. One of the pilots has been a program in Silicon Valley sponsored by Smart Valley, a consortium of local high-tech manufacturers and vendors. It studied 184 employees from Hewlett-Packard, 3Com, Silicon Graphics, and Cisco, as well as a law firm, a consultant, an accounting firm, and Pacific Bell, the local telephone provider. The employees telecommuted at least once a week.

Participants were surveyed before and after the pilot program, and they ranked their benefits at 4.3 on a scale of 5, while concerns ranked 2.4. The benefits exceeded initial expectations in every category: reduced commute time, better quality of personal life, flexible work hours, better quality of work life, improved productivity, attention to environmental concerns, reduced cost of commuting, and flexibility on dependent care.

Their employers also benefited: Telecommuters believed their productivity increased by 25 percent on the days they worked at home. The participants, however, were concerned about their image at work with managers and fellow workers. At least one abandoned the program for fear that he would be out of mind when promotions were due. On the other hand, no problem was perceived in regard to communications between manager and employee. Since the completion of the pilot, many companies in Silicon Valley have undertaken at least some telecommuting activities, usually on an individual basis, such as recent mothers, people with medical problems, and those who have moved with spouses.

AT&T has an aggressive program to allow people to work at home. Currently, 30,000 of its employees nationally telecommute from home on a regular basis under its Alternative Officing program. The company has released results of a cost-benefit study it conducted at its northcentral New Jersey site, where 600 employees have been telecommuting over a five-year period.

The biggest benefit to AT&T was in real estate costs. Allowing employees to telecommute enabled the company to close a whole office complex, a savings of more than $6 million per year. Based on interviews, AT&T also believed that employees worked two-and-a-half hours more per week. That was a gain of $5 million. Greater productivity accounted for $3 million, primarily due to fewer interruptions. Balancing that the first year were estimated costs per employee of $3,000 for office alteration and $4,000 for computer and phone installations. These costs were depreciated over five years, and $1,250 was added for phone, fax, copy, and postage bills, a cost of $3 million per year. The estimated net gain for the 600 employees was $11 million per year after the expenses.[14]

■ TRAINING

Transient staffs, rapid changes in markets, and technology and government regulations are placing a high priority on employee training. Many companies have converted from classes led by staff members to outside consultants and to courses offered on video or, increasingly, via interactive CD-ROMs, kiosks, or the Internet. Of these, the Internet is most attractive since it allows instant upgrading as well as a feedback mechanism that can be used to monitor and document progress, an especially important issue for mandated training.

As happens with all new areas of automation, most corporations had to begin by producing courses with general-purpose graphics, presentation, and Web authoring software, but an increasing number of courses are being developed and offered by specialists. New multimedia authoring programs also make it easy to create specialized courses or customize general courses to specific companies. Technology allows companies to conduct courses long-distance at reasonable cost.

➤ Distance Learning at United Technologies

United Technologies Corporation is working with Boston University to make distance learning a reality by bringing advanced engineering studies into the corporate environment. This virtual university uses PictureTel videoconferencing systems to offer graduate-level courses at four distant United Technologies sites. This approach to instruction maximizes an employee's quality time by eliminating the need to travel to and from the classroom. Because of its interactive capabilities, students have the opportunity to better absorb material. The link between the school and corporations provides enrichment for BU's on-campus students as well. Now more students can experience real-world implementation of textbook concepts. The program also allows UTC to maintain technical competitiveness by recruiting top-notch engineers.

Videoconferencing not only increases employee productivity for UTC but also increases the productivity of the BU professors. By simply overlaying a few courses in the virtual university onto the professor's on-campus classes, the productivity of the professor doubles or triples. The BU program also reaches more than 150 employees at UTC's Hamilton Standard, Sikorsky, Pratt & Whitney, Otis, Carrier, Automotive, and Norden subsidiaries.

BU provides access to faculty for one-on-one time for students, expanding office hours, as well as for project reviews, seminars between BU and corporate customers, and recruiting.[15]

➤ Safety Training with Video

Lam Research Corporation in Fremont, California, is a global company that produces semiconductor manufacturing equipment in both the United States and Korea, and has sales and service offices worldwide. Only 15 years old, Lam is one of the fastest growing companies in its industry and has annual sales of more than $1 billion. Lam uses videoconferencing to ensure that Lam employees get the training necessary to prevent accidents. Videoconferencing lets it train new employees worldwide. In addition, with videoconferencing as a training tool, Lam can be in compliance with Occupational Safety and Health Association (OSHA) regulations. It's also proven to be an effective medium for showing slides, overheads, videos, and white-boards, and for demonstrating specific safety equipment. According to the firm, video reduced travel by 30 trips last year, but, more important, it helps to keep employees from getting hurt.[16]

■ TECHNOLOGY AND ORGANIZATIONAL DEVELOPMENT

Just as new automobiles shut down if their microprocessor brains die, the modern corporation is totally dependent on excellent communications among all its internal parts—and its customers and suppliers. As corporations evolve toward a new organization based on actual work process instead of traditional functions, they must continue to improve their lines of communication until they become fully integrated and able to communicate in real time. At the same time, the "contract" that has governed relations between corporations and their employees has been undergoing a radical change as the corporations reengineer to focus on core tasks and improve productivity. This has had a big impact on employees: Many have been displaced, some to become contractors or work for contractors, but with those remaining considered more valuable. Both groups are benefiting from improved communications, which tends to expand staff boundaries beyond their narrow departmental interests. The Internet, in its internal intranet form, is clearly the most important development in corporate communications since electronic mail first found wide use.

The Internet, and more conventional teleconferencing technology, is also having a big impact, allowing people to work together more effectively, even if they're far apart. These capabilities are also making telecommuting practical for many workers who could never have worked at home or away from corporate centers in the past. Many companies allow or encourage at least some telecommuting, and it's sure to mushroom as employers learn its benefits and become comfortable with remote employees.

Chapter 10

The Corporate Root Canal: Implementing New Technology

■ INTRODUCTION

Choosing any new technology for a corporation—or even choosing whether to use a particular technology—is a scary proposition. It's expensive; it disrupts existing operations; it may not work. There are too many choices and they change too rapidly. Business and trade magazines are filled with horror stories about technological disasters, but they're also filled with examples of successful implementations that have had very beneficial effects, like most of those in this book.

Choosing a technology is intimidating enough if it doesn't change the way your corporation works, but it's terrifying if it forces a complete change in your processes and organization. It's the corporate equivalent of a root canal; everyone dreads it, and it's painful for many involved, but it benefits the organization in the long run, sometimes dramatically.

An example of a technology that totally changes the way organizations operate is the integrated enterprise applications offered by companies like SAP and PeopleSoft. Generally sold to the highest levels of corporations over a year-long sales cycle, the multimillion-dollar costs of these systems pale in comparison to the potential consequences if they don't work.

Fortunately, that's unlikely to happen. What is likely is more grief and expense than anyone expected, and it takes a confident CIO or CEO not to sleep lightly as he or she waits for the system to go live.

Many consultants wait to help corporations choose and install these systems, and they can help with every stage of the process, from deciding what makes sense to the final coding and training. What they all start with, however, is trying to understand your business. It's usually a major project.

Consultants come in two general classes: those who only give advice and those who actually implement systems. Gideon Gartner is among the most respected of the former among information technology users and suppliers alike. Gartner started his Gartner Group to help computer users know what IBM was planning at a time when it was really the only computer company that mattered. Over time, of course, the technology business changed, and the Gartner Group grew into the largest advisor to computer users, giving them unbiased advice about suppliers, trends, and technologies. He sold the Gartner Group years ago and now is running Giga Information Group, which is itself using the new technology of the Internet to deliver advice to corporate users.

Consulting firms like Giga are sitting high these days. Technology has become so important—and so confusing—that most corporations subscribe to counseling services that provide objective looks at technology and the future. Likewise, hardly any company starts a major project without hiring experts to help it decide what to do. Unlike intermediaries who worry that technology may render them redundant, these counseling organizations have moved above the fray, where they can collect healthy fees for their knowledge and opinions.

Gartner considers his firm a "decision support company," in imitation of the decision support systems companies establish to help them make the right choices. "There's a fundamental divergence between the geometrically increasing volumes of information, and our finite human absorption capabilities. The information explosion—perhaps it should be called the misinformation explosion—is flooding us with unmanageable information. To deal with it, we need help."

He lists a number of ways to deal with this data—some of which his company stands ready to provide:

➤ Search engines (though they have a long way to go)

➤ Communities of interest tied to these search engines to exchange expertise

➤ Navigational technologies that go beyond simple searches

➤ Abstracting and synthesizing engines

➤ Human intervention by dedicated advisors with diagnostic capabilities

➤ Calibration of sources to normalize against biases

➤ Graphic and network user interfaces to underlying technologies for authoring and display, for push and pull, which are still evolving

➤ Evolution of research from the static, publication model to dynamic, living research

■ THE FIRST STEP

Every company is unique, and every one has different needs. This applies to the technology a company uses as much as to its products, production, and distribution methods. For that reason, the first step in deciding what technology to use is, ironically, to step back and instead look at your company and its operations. It's not just meaningless talk to determine what your company is and what it wants to do. Not only does technology follow processes, but process must follow technology.

The most basic question is corporate strategy. Amazingly few corporations really have articulated their strategy; they usually create fuzzy phrases designed for marketing purposes, not for driving operations.

There are only a few basic corporate strategies—notably, innovation, being the low-cost supplier, focusing on a niche or customer service. Each company must pick one as its primary strategy, because these can be mutually exclusive or even conflicting as fundamental strategies. But the wise company will monitor its role, choosing where to devote secondary effort. It can't overlook the threats it may face from other companies that enter its markets with other strategies.

In computers, for example, many companies such as IBM and Hewlett-Packard want to think of themselves as innovators, and they have certainly succeeded in creating many significant new products and even markets. Fundamentally, however, their strengths have been customer service and support, allowing them to charge more to customers who are willing to trade a higher price for the peace of mind that comes with buying high-quality products from a supplier who will stand behind them.

Dell and Gateway, by contrast, built healthy businesses by being low-cost suppliers. Following direct-sales patterns well established in other markets, they appealed to the buyer for whom price is all-important. However, Intel is a true innovator, creating products and market demand that are basically immune from price pressure and customer service.

All these companies have succeeded, but all have stumbled when they strayed into other strategic directions. Intel, for one, has repeatedly tried to enter consumer markets with board-level peripherals and other products where it ultimately couldn't compete. And with rare exceptions, Hewlett-Packard hasn't succeeded when it tried to innovate in a market that preferred the comfort of evolution and support. The first to introduce a true portable computer, it deviated from existing industry standards, as it also did when it chose a nonstandard $3\frac{1}{2}$-inch diskette for its desktop computers instead of the $5\frac{1}{4}$-inch floppy disks customers then preferred.

■ STRATEGY BEGETS PROCESS

Once a company understands its priorities, it can plot its processes with the help of consultants and vendor experts. In some cases, companies at this juncture make significant changes in organization, even shedding nonstrategic operations or completely changing responsibilities of executives and their organizations. Then the experts design processes, matching them to software and hardware to create a system that accomplishes what the corporation wants and needs.

For enterprise systems, there are two choices: the so-called big bang, and a function-by-function rollout. Completely changing all a company's processes and systems at once is bound to be traumatic, but for many observers, the slower approach is more disruptive in the long run.

The ideal situation, of course, would be to start from the beginning with a complete system, but few companies have that luxury. Even well-funded start-ups rarely adopt integrated systems because of the expense and complications. Some firms have been able to begin with a clean slate, and it creates major benefits. There may also be complications, however, particularly if the new company spins off from another one.

➤ An Instant Company with an Instant IT Operation

Cable Systems International (CSI) is a $400 million manufacturer of copper cable for the telephony industry. CSI began life in October 1995 when seven former AT&T employees put together financing and bought the operation from the telephone giant. Its customers are mainly regional Bell operating companies (RBOCs).

CSI had an unusual opportunity and a challenge as a result of starting fresh as a relatively large firm. It went live with SAP's R/3 integrated software the same day it started life as a new company. CSI's bank required it to have computer systems in place to handle its business operations before it would provide the necessary loan. By necessity, CSI took a big bang approach, going live to 384 users in 14 locations throughout North America. Later it added a sub-assembly plant in Mexico. Andersen Consulting assisted CSI in the process.

Prior to CSI's formation, it relied on AT&T's IT systems, but the data it needed to run its own company resided on some 77 different legacy systems throughout AT&T. It took 18 months for implementation time, and getting data off AT&T's 77 disparate systems was a big part of the job.

The company uses many SAP modules, including the WM warehouse management that is used to track consignment and inventory and to archive data. They run on Hewlett-Packard computers and an Oracle database. The 384 users (210 of them concurrently) use PCs.

CSI exceeded its business plan's projections for first-year revenues, making its financial goal in only 11 months instead of the planned 12. Vice President of Operations Tony Piccirillo says R/3 contributed directly to this by helping the firm collect receivables earlier because the real-time access means invoices get out quickly so receivables are booked sooner. This has put CSI in an excellent cash-flow position and it hasn't had to borrow money again to further finance and grow the company. The company has also reduced inventory by about 20 percent from what it inherited from AT&T, mostly attributable to better forecasting and planning enabled by R/3. Cable Systems employs EDI for supply chain management and fast, efficient receipt and processing of orders.

Cable Systems had to hire 35 new IT people with client/server and UNIX experience to create its information systems. Most of the MIS employees in the AT&T division that spawned Cable Systems had host-based, mainframe backgrounds that weren't going to transfer well to the new company's client/server systems. Many of these people instead chose AT&T's voluntary retirement program. It was imperative for Cable Systems to hit the ground running, so it hired new people with expertise rather than investing in the career retooling these people would have needed. Once the system was installed, Cable Systems did its own R/3 training: IT trained 20 "power users" who then trained the other 350 users over a six-month period.

Overall, Piccirillo feels R/3's biggest asset is that it's a "user system" not an "MIS system," which is what he had encountered at AT&T. Batch processing, multiple databases, and lots of custom programming typify most non-SAP environments, adding overhead to IT administration because it generally takes a programmer to get any real data out of the system. By contrast, SAP is user-driven and user-accessible.[1]

➤ A Merger and Spin-off

Like Cable Systems, Cultor Food Science started big, as a combination of spin-offs and an acquisition. CFS of New York is a $500 million division of Cultor Limited, a $2.2 billion Helsinki, Finland, food and nutrition company. Cultor Food Science was formed in January 1996 when Cultor acquired the Food Science Group from Pfizer and combined it with the Flavoring and Xyrofin divisions of Cultor Limited. Cultor Food Science, developers and marketers of unique high-performance food ingredients, focuses on specific food industry needs: flavorings, protection, and nutritional enhancement.

As a newly formed organization, Cultor Food Science faced a need to migrate off of the IS systems of its former parent. North American offices are on SAP now; international rollout should be completed by July 1998. SAP Consulting Services handled the implementation under its quick-turnaround ASAP Program.

The firm uses nine SAP R/3 modules. The server platform is a Hewlett-Packard HP 9000 running an Oracle database. It took six months to go live to all 150 users in North America, though three modules were implemented a few months later and international rollout was completed within a year, bringing the total number of users to approximately 250.

CFS's long-term strategy for leveraging IT for competitive advantage could not be met by a mainframe system. CFS needed a scalable client/server solution that would support better decision making through an integrated view of corporate data. R/3 was the best fit for CFS's business requirements without requiring customization; it found R/3 to be inherently well suited to its business processes. CFS expects to improve customer service by attaining a real-time, real-world view of all customer requirements and data, allowing it to adjust more dynamically to changes in requirements and to amass—and use—a database with marketing intelligence on the customer's profile.

A consolidated view of customer data is mission-critical for CFS. Many of its products are customer-specific. Customers may change requirements such as the degree of sweetness, viscosity, or granularity, and CFS must be able to respond, ideally without delaying time to market.

CFS estimates it saved in excess of $2 million the first year as a result of more accurate asset management, better inventory control (reduce time in-stock), and reduced time to market through least-cost distribution schemes, enabled by better production planning and delivery schedules. International orders can now be processed the same day as received. In the old environment, this process would take three days because of batch links between systems.

The SAP team for CFS numbered 16 people, while CFS's internal team included 10 full-time and 5 to 10 part-time members. Initially, SAP team members interviewed users and conducted a need assessment, which was then mapped to a specific implementation plan for R/3. The consultants are experts in particular modules of R/3 and specialize in particular business processes.

The full-time members of the CFS core team were drawn from management and empowered to make all decisions, with the explicit commitment to resolve any issue within 48 hours. A steering committee of senior management, representing all affected business functions, met monthly to review progress, and committee members were individually available on request to deal with issues needing escalation and resolution. An IS team of three CFS people was committed full-time to the project and given the job of creating the needed infrastructure, such as the hardware and wide area network needed for the project.[2]

➤ Replacing an Obsolescent Orphan

A company's computer system should meet its needs and provide it with a competitive advantage in its business environment. With an outdated system, its sales forecasts lag behind the market, its raw material inventories can't keep up with manufacturing, and delayed billing and distribution systems can strangle its cash flow.

These are just some of the problems New York–based Forest Pharmaceuticals faced when Qantel Computer went out of business in the early 1990s. Forest chose a Qantel system in 1986 with no idea it would become obsolete so soon.

The parent company of Forest Pharmaceuticals, Forest Laboratories, develops, manufactures, and sells both branded and generic forms of ethical drug products for the treatment of a wide range of illnesses.

From its humble beginnings 20 years ago as a vitamin, ice cream, and candy company with $5 million in yearly sales, Forest Labs more than doubled its growth in the past 5 years and reported revenues of more than $400 million in 1995.

Forest Labs produces and markets pharmaceutical products that are important to the health of Americans. These include Aerobid, an inhaled steroid treatment for asthma; Flumadine, which reduces and treats flu; Monurol, a single-dose antibiotic for urinary tract infections; and Cervidal, which eases inducement of labor.

A pharmaceutical company that stands still will soon be left behind. Not only must an industry competitor focus on existing products and short-term opportunities, but it must also identify new product opportunities. Startling scientific advances will be produced in the coming decades, including the ability to affect genetic activity in the nucleus, control the inflammatory process, regulate the immune system, replicate tissue and neurons, and deal with aging and cancer. The rapidly changing opportunities demand that pharmaceutical companies must use every advantage at their disposal to remain competitive. That includes their computer systems.

Forest's existing proprietary Qantel system didn't interface with other systems and its technology was obsolete. There was no off-the-shelf software available for the computer, so Forest programmers had to write special applications and the company had to rely on a collection of applications to interface between various business functions.

Redesigning a computer system for a company presents a challenge for even the best-organized company, and a team of four top-level Forest managers from manufacturing, sales and distribution, financial, and IT spent a year evaluating new hardware and software before arriving at their decision. After their experience with proprietary systems, they thoroughly embraced standards: UNIX, open systems, interoperability, and a Windows-based interface.

The software came down to several choices, which were evaluated for their ability to offer an enterprise-wide business solution. One software package was ruled out because the company was small and the selection team correctly surmised that it wouldn't remain in business over the long run. Another was ruled out because its product interfaced with only one database.

On the other hand, SAP R/3, from SAP America, is a feature-rich package that is highly configurable and provides all of the business functions needed by Forest Pharmaceuticals. No major modifications were required. The fact that companies like Microsoft, Informix, Anheuser-Busch, and Monsanto were signing up with SAP helped Forest make the decision. SAP also worked with the Informix database that the company used for limited functions.

Implementation of SAP on Sun equipment began in August 1994, and the manufacturing and production modules were installed a year later. The company continues to expand the system to other operations and upgrade to new functions.

Forest Pharmaceuticals implemented most SAP modules—manufacturing, production, sales and distribution, financial, and costing, virtually every function needed to run a business except human resources—without major software modifications. The only changes it had to make were based on government regulations for the pharmaceutical industry. The company, unlike many, did not have to undergo business process reengineering, because Forest Pharmaceuticals is already lean and efficiently run.

Now that fears of facing the future with an antiquated computer system are behind it, Forest Pharmaceuticals is concentrating on further implementation and integration of this enterprise-wide answer to running its business. The SAP R/3 software, with its ability to utilize information from the Informix data warehouse, creates a comprehensive solution in keeping with the company's efforts to operate a lean, efficient organization. Sun provides flexibility, interoperability, and a clear upgrade path that will support Forest Pharmaceuticals' growth plans well into the twenty-first century.[3]

➤ The Right Product with the Right Team

Unlike Forest, CSI, or CRS, Microsoft replaced existing systems with an integrated enterprise system. The process was an intense one, partly because of Microsoft's fast growth and partly because it felt it had to make a showcase implementation of the product on its own technology. Ignoring this requirement, the experience is a textbook example of installing enterprise software.

Microsoft's decision to implement SAP R/3 was not taken lightly. Twice before—first in 1992 and later in 1993—the company evaluated a number of software packages including SAP R/3 and decided not to make a change at that time.

Microsoft finally decided it had to improve its systems. It selected SAP R/3 because of its functionality and global scalability. Microsoft needed a sophisticated product that could handle a complex multinational structure. It also, not surprisingly, insisted on a company committed to implementing its product on Windows NT and SQL Server to take advantage of the open client/server architecture, homogeneous operating system, and integration within the Microsoft BackOffice family, as well as with Microsoft Office products like Excel and Word.

Like its evaluation of software vendors, Microsoft also spent a lot of effort choosing a consulting partner to help it install the product. It chose Deloitte & Touche Consulting Group/ICS. According to Microsoft Project Manager Aidan Waine, "D&TCG/ICS had the strongest overall team by a fair margin. Besides its FastTrack 4SAP methodology designed to reduce the implementation's time frame, we were really impressed by their aggressiveness and proven ability to get clients up and running in record time."

Microsoft committed equal numbers of business and information technology staff to the project, providing a solid balance of business needs with technical support knowledge. This shared responsibility strengthened the involvement and support the team received from executive management, and helped the team make key business process changes quickly, a key to the project's success.

Microsoft kicked off the first phase to implement SAP R/3's procurement, accounts payable, asset management, and finance/controlling master data modules for 100 users at corporate headquarters in September 1995. This phase went live seven months later.

Phase 2's focus was bringing up the general ledger modules on July 1, 1996, for the new fiscal year. This raised the total number of users to 1,300 and completed the transition of the financial structure to SAP R/3. With its first implementation going live on July 1, 1996, in Canada, phase three is the worldwide rollout for 25 additional subsidiaries. At the end of phase 3, Microsoft will be the largest Windows NT and SQL Server SAP R/3 implementation in the world.

The SAP project management group, organized early in the process by Microsoft, was co-led by project managers Gregg Harmon from Microsoft Finance, Aidan Waine from Microsoft Information Technology, and Liz Fasciana from D&TCG/ICS Consulting. Each phase team was also led by a similar trio.

Throughout the implementation, as the project team grew from 20 to 60 members, Microsoft maintained balance between D&TCG/ICS, Information Technology, and Finance. In addition, SAP provided strong project support with on-site involvement and 24-hour phone support.

"The tight integration of the team members brought incredible value to the design effort, and the additional expertise from SAP added to the knowledge pool. These were definitely the keys to successfully meeting Microsoft's needs in this time frame," noted Fasciana.

Microsoft established clear goals for its business transformation process: global financial standardization and streamlining of procurement processes. To reach these high-level goals, Microsoft took advantage of the power of SAP R/3 software. According to Harmon, "We didn't do a 'blue sky' redesign. We worked with the firm's business process reengineering services from the start, achieving process integration by following SAP's natural flow. We allowed SAP to drive the business transformation instead of the other way around."

For Microsoft, the biggest challenge was understanding the details of SAP R/3 and making timely decisions about how to implement business policies. On this point, Harmon explains, "You can either make decisions quickly based on incomplete information or you can try to research every option in great detail. If you go to either extreme you're going to lose momentum—striking a balance can be a challenge." It was within this dilemma that D&TCG/ICS added the most value to Microsoft's efforts. Harmon explains, "The firm was instrumental in helping the team shave down the options. That's the kind of know-how that can only come from experience."

The team completed phases 1 and 2 on time and on budget, strictly adhering to deliverables and milestone dates. To further ensure that schedules and budgets remained on target, the team guarded against "scope creep," a phenomenon that occurs when customers see how robust SAP software actually is. To achieve success, requests to extend the scope of the project had to be evaluated very critically and were not approved if they slowed or redirected the implementation schedule. In circumstances under which it did make sense to change the scope, the team made sure they had the resources to deliver or that the business users got involved to provide the necessary support.

The implementation provided an excellent showcase for Microsoft's application development technology. Microsoft's Object Linking and Embedding (OLE) automation is used to update SAP from Microsoft Excel using Visual Basic for Applications. According to Waine, "It was relatively straightforward to create sophisticated data validation and upload tools with our existing desktop tool set and SAP's integration facilities."

For example, Microsoft works with several strategic vendors that are not EDI-capable. These vendors E-mail their invoices to Microsoft Accounts Payable as formatted Microsoft Excel spreadsheets including Visual Basic for Applications macros, which validate and upload the data to SAP, eliminating rekeying of data.

According to Waine, Microsoft employs the same mechanism to interface financial transactions from other internal systems.

"We still need to post journal entries from about 50 ancillary internal processes and systems that have not converted to SAP. Over 70% of our transactions are entered into SAP through Microsoft Excel interfaces, increasing data accuracy and efficiency."

Perhaps the most compelling of these solutions is Microsoft Market, an Internet Explorer– and SQL Server–based application that allows Microsoft's 20,000-plus employees to submit purchase requisitions through the corporate Intranet. Purchases from strategic vendors are made simple through the shopping basket metaphor. Links to SAP R/3 ensure that the user codes the requisition correctly. Finally, the application uses work-flow logic to route a Microsoft Exchange E-mail message to the appropriate approval authority for tacit approval before transmitting the requisition directly to the vendor. Only requisitions that require procurement intervention are routed into SAP. Besides transaction accuracy and speed, an added advantage over paper-based requisitions is the user's ability to inquire about the status of requisitions through the Microsoft Market tool.

In preparation for the rollouts, the project team had dedicated resources focused on communication, training, and end-user support. The business changes affected most Microsoft employees as a result of the new SAP code structures and reporting tools. This made the implementation issues broader than the functional focus of the project. Centralizing the strategy and implementation for these areas provided a support structure for each of the phase teams and assured consistent and comprehensive delivery. "It was critical that communication, training and support were all handled well as they build off one another and provide an overall network to keep end user productivity as high as possible during implementation," pointed out Harmon.

The training team's objective was to integrate the overall business with the application, covering processes within and outside of the SAP R/3 system. This proved challenging due to the narrow window between final decisions on the system's configuration and the SAP live date. Microsoft used D&TCG/ICS training consultants to design and create the training materials and Microsoft business experts to deliver the classes. According to Harmon, "D&TCG/ICS' training team did a first-rate job developing excellent training materials under intense deadline pressure. The materials were so clear and well organized that our business people were able to deliver the courses with ease and proficiency."

Microsoft set up a dedicated SAP help desk for initial application support, which was involved in the project from the beginning. The structure follows the standard Microsoft internal three-tiered support model and will roll into the broad support group six months postlaunch. The help desk uses the intranet web structure to provide information to users, as well as communicating across the end-user mail alias.[4]

■ IMPLEMENTING TECHNOLOGY

Few companies acquire new technology the way consumers do—just because they want the newest. They take on the cost and disruption because they see a problem that needs to be solved—typically that of being more competitive.

The biggest change, but the one with the greatest potential, is converting corporate processes and systems to an integrated enterprise system of client/server software. Almost every progressive company is either adopting or considering this technology, and, though almost all of the ones that have been through the process acknowledge its pain, they all feel the gain is worthwhile.

Fortunately, there are corporate coaches to ease the process.

Technology Has Transformed Corporations

Even a casual observer can see that technology is having an enormous impact on corporations and the way they operate. Hundreds of computer applications have helped corporate employees operate more efficiently and productively. What is not so apparent to the casual observer, however, is the change in the ways technology is affecting the corporations themselves.

Early applications of technology, starting with the use of computers for accounting and maintaining mailing lists, dramatically improved the productivity of the individuals in corporations. Other dramatic improvements followed, as word processors, spreadsheets, automatic drilling machines, and bar code readers let each individual in the company perform his or her task more quickly, more accurately, and with less repetitive drudgery.

These changes, however, didn't really change the basic organization and operation of companies. They still maintained the same functional breakdown and hierarchical structure created a century ago. Many observers saw that this was not the best way to operate, but proposed alternatives were simply too complex for the existing communications and manageability of the firms. There was no practical way to take all the individual transactions and information that occur in a modern organization and create a unified whole.

Modern information technology has put that goal within reach, however. Using today's distributed client/server computer organization, advanced networking, and applications built on sophisticated relational database managers, it's possible to truly unify a whole corporation, its customers, and vendors. Now every part of an organization—manufacturing, procurement, marketing, finance, product development—can feel the effect of an individual order booked by a sales rep or even a retailer. This allows the company to deliver individual products quickly with minimal inventory, even as it's collecting data that should help it design its next promotion and future new products.

In the process, these integrated enterprise systems are fundamentally changing corporate structure and operation. Corporations can finally focus on their strengths without having to compromise because of existing overhead and traditions.

Many companies have already spun off nonstrategic activities to specialists for whom those activities are strategic. Most noted are closing and selling manufacturing plants and subcontracting for assemblies or whole products. If a company has an expensive automated factory designed to make circuit boards containing through-hole parts, it isn't as likely to jump to new surface-mounted parts, even if that's what economics and customers prefer. If the company uses suppliers that focus on manufacturing, however, it's likely to be able to immediately respond to market and technology trends—and its supplier is likely to be ready to respond, too, or else to suffer. These suppliers, however, can no longer be considered simply vendors. Many must become true partners, sharing information openly—another built-in feature of today's enterprise resource systems.

There are many other examples of outsourcing as well. These include everything from receptionists and security to employee cafeterias, warehousing, and transportation—activities most companies once considered necessary parts of their being.

Just starting, but sure to be a big trend, is automating and farming out human resources functions, a complex, expensive, and generally unrewarding headache. The Internet and its siblings, the intranet for employees and the extranet for customers, suppliers, and partners, are emerging as key catalysts for these and other changes.

These changes are helping companies become leaner, increasing their return on investment and profitability while allowing them to move more nimbly in fast-changing markets. For employees, these trends can provide more rewarding and satisfying careers. Many jobs involve repetition and drudgery, and an integrated corporate information system can eliminate much of this boring work. New systems can create the opportunity for employees to be more creative, to better demonstrate their value, or even to work at home part- or full-time if that's their desire.

Another benefit to both corporations and their employees is that today's networking, intranets, and integrated information systems tend to flatten organizations, reducing the need for layers of unproductive middle managers. This also helps the organizations become more efficient and productive, while increasing job satisfaction among the staff.

Corporate changes can also lead to fewer jobs. For the ambitious and educated, these changes are likely to be positive, but they can also force employees to accept jobs with less pay or fewer benefits—perhaps doing the same job for suppliers to their former employers.

■ IMPLEMENTING TECHNOLOGY

One of the biggest changes that has occurred in implementing technology is that more and more companies buy software applications rather than writing their own. This is true because more suppliers are responding to the need for both enterprise-wide and specific applications, and because most companies are realizing that their resources are best spent focusing on their own businesses, not developing tools.

With this change comes a new challenge: selecting the right applications vendors. These vendors, in fact, must become partners, because the process of configuring, installing, and troubleshooting software, plus long-term support, become critical when a corporation's whole operations are tied to a single information system. Suppliers of large enterprise packages have developed sophisticated, long-term sales programs aimed at corporate executives, not information technologists, to help them sell these expensive but critical packages. One result of the importance of this decision process and the implementation is that most companies select consultants to help them through the process—both the process of choosing the best package and the implementation.

Often, however, the same consultants who attempt to provide objective advice about which applications vendors and software to choose also act as systems integrators to install and customize the software. Some corporations choose different advisors for the selection process and the implementation; others choose consultants who work with many software vendors, hoping to reduce the bias and increase expertise in that way.

Whatever help is chosen, it usually takes a lot of work and time to adapt the software to a company's operation—or, more likely, to change the corporation itself to operate more effectively using the software. Many problems in installing the applications and coming on-line really result from the changes normally required at corporations that choose to reengineer operations and implement a new information system simultaneously.

■ POWER FROM INFORMATION

Whatever the pain involved in adopting enterprise-wide information systems, companies that have switched report dramatic results. Hewlett-Packard, for example, has reduced its sales, general, and administrative (SG&A) costs from 28 to 17 percent between 1990 and 1996, partly due to its use of SAP R/3 software, says chairman Lewis Platt.

Many users of SAP R/3 software mention specific benefits in a report in *Business Week* (May 19, 1997). Owens Corning, which serves the flat market for glass and insulation products, cannot raise prices, notes CIO Michael Radcliff, so the firm invested $175 million to revamp its computer systems and link its 150 operating locations. He expects the system to deliver an annual boost to productivity and has already seen results: The firm's earnings jumped 8 percent in the first quarter of 1997 on just a 3 percent rise in sales.

Whirlpool also reports that new equipment and methods are delivering productivity gains to the appliance maker. Its earnings rose 21 percent in the first quarter of 1997 despite a 1 percent decline in revenues. One reason is that a new factory with design and manufacturing improvements can deliver products quickly and in bigger volumes with fewer parts, partly due to SAP software.

IBM currently has 21 SAP projects under way, covering 80 percent of its core business activity; 8 are already running and producing results. At its large disk-drive operation, for example, before the firm installed R/3, credit for returns took three weeks; now it occurs on the phone when the customer calls. Before R/3, it took thousands of hours of gathering and reconciling data from systems to prepare end-of-month reports. Now the process is automatic. IBM CIO Gerald Prothro said "[IBM] liked the idea that the best practices in business were embedded in the SAP software."

Microsoft saves $12 million a year just in early payment discounts from vendors, thanks to its worldwide common procurement system. It saves $2 million in equipment depreciation alone, says CIO John Connors. Before installing the SAP software, the company took three months to start a depreciation schedule for a newly purchased asset; now it is done immediately. "When we started putting some of these benefits before Bill Gates, his eyes got wide," said Connors.

Colgate-Palmolive, with $8 billion in sales and 30,000 employees worldwide, installed an integrated supply chain using R/3 to help support corporate strategy, said CIO Bruce Johnson. It is designed to provide superior customer service through fast, frequent product deliveries, enabling rapid response to changes in demand. Johnson says the R/3 software has helped to reduce order-to-delivery cycles, manufacturing cycle times, closing times, and working capital, and has streamlined the company's overall cost structure. The firm has been able to consolidate more than 70 data centers to just a few with a global infrastructure. R/3 provides complete integration across the entire supply chain, automating the order path from customer request through fulfillment. The system also provides better service levels, a higher percentage of on-time completions, and an improved match between what is ordered, what is delivered, what is invoiced, and what is paid.

At Computervision, an ROI study showed that with 10 percent increased volume, half the budget, and 30 percent of the head count, the software provided substantial improvement in many areas. The time it took from quote to cash dropped from 270 days to 65 to 80 days, the time between quote and invoice fell from 59 to 67 days to 2 days, the order-entry time went from 6 days to less than 1 because it was automated, and financials close in 5 to 7 days instead of 14 to 16 days. The number of invoices with errors has declined from 18 percent to 10 percent, and the accounts receivable over 90 days have dropped from 18 to 25 percent to 9 to 14 percent.

At Fujitsu Microelectronics, cycle time for quotations dropped 90 percent, from 20 days to 2 days, for 60 to 85 percent improvement in on-time delivery, and 50 percent reduction for financial closing times, from 10 to 5 days.

The Boston Beer Company was able to reduce the time it took to gather inventory data from suppliers from four weeks to one week. It also was able, for the first time, to perform profitability analysis and improve its forecasting. At Boston Beer, the R/3 administration involves only a three-person IS team.

Pharmaceutical maker Bristol-Myers Supply now gathers detailed information on how doctors prescribe drugs, so its sales reps can be far more productive. Chief executive Charles A. Heimbold Jr. sums up the impact of new integrated enterprise systems, "We will look back to this as a time when companies' investments in computing power and software started to pay off."

Users of other enterprise software products report similar results. The evidence is clear. After 50 years, modern integrated computer information systems are really transforming corporations, not just jobs. Those corporations that fully embrace their possibilities have a chance to lead in the future; those that don't are sure to end up in the backwaters of our economy.

Appendix

■ A WORD ABOUT TERMS

People who work in information technology, like all specialists, have their own vocabulary. Some terms come from the simple necessity of having names for things that these people use. Others come from the need to be very precise or, ironically, very vague, about concepts that are unique to their businesses. And some seem to serve primarily to establish a community and exclude others. You'll never hear an MIS person refer to a "computer," for example, but to "hardware." And a platform isn't what the hardware sits on, but is the basic software environment underlying applications such as accounting or procurement. Whatever the reason, it's something that the outsider must deal with in order to understand and communicate with IT specialists.

In this book, I've tried to avoid acronyms and true jargon as much as possible, but you obviously can't avoid using specialized terms if you're writing a book about technology. The following terms occur a great deal in the text, however, and here's a simple explanation of what they usually mean.

client   A workstation in client/server computing.

client/server computing   Computer systems that distribute work between workstations (clients) and servers (such as database servers).

data   Pieces of information stored in a computer. They can be numbers, text, pictures, or something even more complicated.

database   A computer file containing data—usually lots of it. Most modern database managers are relational; that is, they can maintain links between individual data in the records.

decision support systems   Programs that help company managers choose between options. Most include programs that simulate and optimize processes and programs, and may include expert systems that try to suggest approaches based on knowledge collected from experts.

enterprise   A whole corporation or other organization.

hardware   Computers and other physical devices such as printers and networking equipment.

infrastructure   Everything involved in or part of a project or situation.

legacy   Existing entities, particularly referring to old mainframe computers and the programs and data they contain.

mainframe   A big computer, usually made by IBM.

MIS   Management information systems; an old term for what is now generally called *information technology* or *systems*. Prior to being called MIS, it was called EDP (electronic data processing).

objects (as in "object-oriented")   Self-sufficient software modules that can be combined to create programs that solve larger problems. Objects can be data or programs.

platform   An underlying software environment (sometimes specific hardware).

programs   The software that tells computers what to do.

reengineering   To reorganize an existing company from the ground up to make it operate more efficiently.

server   The computer that manages data files or serves another specialized use. It may be what used to be proudly called a *mainframe* or *minicomputer,* or it may be a souped-up personal computer.

solution   A program designed to solve a big problem.

work flow   The way information is processed by people and organizations. Usually used in the context of work-flow automation programs that encourage or force people to do things more efficiently.

workstation   The computer that an individual uses. It can be a personal computer or an engineering-oriented computer from a company like Sun or Silicon Graphics.

Notes

■ INTRODUCTION

1. Louis V. Gerstner (keynote speech presented at the Gartner Symposium, Orlando, Fla., October 9, 1991).
2. Gerstner, speech, 1996.
3. William Gates, "The Digital Nervous System" (keynote speech presented at the 1997 Microsoft CEO Summit, Seattle, Wash., May 8, 1997).
4. Gerstner, speech, 1996.
5. Gates, speech, 1997.
6. Lewis Platt, "Creating the Extended Enterprise" (presented at the net.profit Executive Summit, Geneva, Switzerland, February 7, 1997).
7. Gates, speech, 1997.

■ CHAPTER 1

1. IBM CATIA Web site, http://www.catia.ibm.com/custsucc/supfal.html.
2. IBM CATIA Web site, http://www.catia.ibm.com/custsucc/susamp.html.

3. IBM CATIA Web site, http://www.catia.ibm.com/custsucc/sufrig.html.

4. IBM CATIA Web site, http://www.catia.ibm.com/custsucc/subla3.html.

5. Silicon Graphics Web site, http://www.sgi.com/works/successstories/marine.html.

6. Smithsonian Innovation Entry, http://innovate.si.edu/x/view.pl?nomid=97196.

7. IBM CATIA Web site, http://www.catia.ibm.com/anncpres/acchry.html.

8. Silicon Graphics Web site, http://www.sgi.com/works/successstories/chrysler.html.

9. EDS Unigraphics Web site, http://www.ug.eds.com/publications/success/caterpillar.html.

10. EDS Unigraphics Web site, http://www.ug.eds.com/publications/success/gmemd.html.

11. Azizia DTM Web site, http://azizia.dtm-corp.com:80/kodak-ap.html.

12. Azizia DTM Web site, http://azizia.dtm-corp.com:80/porsche-.html.

13. Azizia DTM Web site, http://azizia.dtm-corp.com:80/hp-appl.html.

14. Cubital Web site, http://www.cubital.com/cubital/nera.html.

15. 3D Systems Web site, http://www.3dsystems.com/library/edge/52feature2.htm.

16. Cadence Web site, http://www.cadence.com/features/archive/vol1no2/motorola.htm.

17. Informix Web site, http://www.informix.com/informix/solution/manufact/2030573/2030573.html.

■ CHAPTER 2

1. Jim Hine, Computer Sciences Web site, http://www.csc.com/aboutcvr_emi.html.

2. Smithsonian Innovation Entry, http://innovate.si.edu/x/view.pl?nomid=96277.

3. Informix Web site, http://www.informix.com/informix/solution/manufact/2069973/2069973.html.

4. SAP Web site, http://www.sap.com/usa/index.htm.
5. SAP Web site, http://www.sap.com/success/ibm.htm.
6. Informix Web site, http://www.informix.com/informix/solution/manufact/2063673/2063673.html.
7. Smithsonian Innovation Entry, http://innovate.si.edu/x/view.pl?nomid=97165.
8. Masi Web site, http://www.masi.com/success7.htm.
9. Universal Instruments Web site, http://www.uic.com/thomson.html.
10. Universal Instruments Web site, http://www.uic.com/genie.html.
11. Annette Shimada, Baan, E-mail to author, 12 March 1997.
12. Lou Harm, Burson-Marstellar, E-mail to author, 10 March 1997.
13. Autodesk Web site, http://www.autodesk.com/solution/customer/toyota.htm.
14. Compaq Web site, http://www.compaq.com/solutions/stories/.

■ CHAPTER 3

1. Robert D. Shecterle, "Global Supply Chain Management," *Oracle* (May/June 1997); available at http://www.oramag.com/archives/37/37apps.html.
2. Smithsonian Innovation Entry, http://innovate.si.edu/x/view.pl?nomid=96458.
3. Manugistics Web page, http://www.manugistics.com/html/skp.html.
4. Manugistics Web page, http://www.manugistics.com/html/bfp.html.
5. "Hewlett-Packard Packs Power into Personnel and Manufacturing Programs" (PeopleSoft document, June 1997).
6. "Objects for Health," *Oracle* (January/February 1997); available at http://www.oramag.com/archives/17/17toc.html.
7. Compaq Web site, http://www.compaq.com/corporate/edi/edi-def.html.
8. Sterling Commerce Web site, http://www.stercomm.com/success/newbal.htm.

9. Sterling Commerce Web site, http://www.stercomm.com/success/cumm.htm.

10. Ibid.

11. DNS Web site, http://www.dnsww.com/case.html.

12. Ibid.

13. Premenos Web site, http://www.premenos.com/news/stories/abbott.html.

14. Premenos Web site, http://www.premenos.com/news/stories/ss11.html.

15. Hewlett-Packard Web site, http://www.hp.com/gsyinternet/stories/59656448.html.

16. "SAP and Aspect Team Up to Reduce Manufacturers' Procurement Costs and Time to Market" (Aspect release, 20 February 1997).

17. Ibid.

18. Alan Ampolsk, Fleishman-Hilliard, E-mail to author, 21 July 1997.

■ CHAPTER 4

1. William Gates, "The Digital Nervous System" (keynote speech presented at the 1997 Microsoft CEO Summit, Seattle, Wash., May 8, 1997).

2. Compaq Web site, http://www.compaq.com/solutions/stories/trueval.html.

3. Zpix Web site, http://zpix.com/casestudies.

4. Informix Web site, http://www.informix.com/informix/solution/manufact/2037473/2037473.html.

5. Freshly Wired Web site, http://freshlywired.com/release2.htm.

6. Compaq Web site, http://www.compaq.com/solutions/stories/intelck.html.

7. Hewlett-Packard Web site, http://www.hp.com/gsyinternet/stories/tandy.html.

8. Tandem Web site, http://www.tandem.com/cust_ref/amertccr/amertccr.pdf.

9. Cognos Web site, http://www.cognos.com/busintell/moen.html.

■ CHAPTER 5

1. William Gates, "The Digital Nervous System" (keynote speech presented at the 1997 Microsoft CEO Summit, Seattle, Wash., May 8, 1997).
2. Diane Carlini, Symantek, E-mail to author, 20 June 1997.
3. Goldmine Web site, http://www.goldminsw.com/media/html/case_studies.htm.
4. Todd Fryburger, "Holiday Inn Worldwide: Field Force Automation and the Evolution of the Hospitality Industry," Computer Sciences Corporation Web site, http://www.csc.com/about/holiday_inn.html, 1995.
5. Trilogy Web site, http://www.trilogy.com/press/pr10196 chrysler.html.
6. Alan Ampolsk, Fleishman-Hilliard, E-mail to author, 18 July 1997.
7. Smithsonian Innovation Entry, http://innovate.si.edu/x/view.pl?nomid=97118.
8. Smithsonian Innovation Entry, http://innovate.si.edu/x/view.pl?nomid=97136.
9. IEX Web site, http://iex.com/new/call_ctr/mci-tv.html.
10. Computer Sciences Corporation Web site, http://www.csc.com/about/news_stories/csc_retail.html.
11. Trilogy Web site, http://trilogy.com/press/pr070997volvo.html.
12. Alan Ampolsk, Fleishman-Hilliard, E-mail to author, 18 July 1997.
13. Hewlett-Packard Web site, http://www.hp.com/gsyinternet/stories/59657912.html.
14. Smithsonian Innovation Entry, http://innovate.si.edu/x/view.pl?nomid=97289.
15. "software.net and Microsoft Ink Largest Electronic Software Distribution Deal in Internet History" (software.net release, 1 July 1997).
16. Jill Donnley Rege, "Oracle Helps Internet Entrepreneurs Bring the Car Lot to the Desktop," Oracle (July/August 1997); available at http://www.oramag.com/columns/jillauto.html.
17. Smithsonian Innovation Entry, http://innovate.si.edu/x/view.pl?nomid=97121.
18. SAP Web site, http://www.sap.com/success/guinness.htm.

19. Tandem Web site, http://www.tandem.com/cust_ref/ vicseccr/vicseccr.pdf.

20. Smithsonian Innovation Entry, http://innovate.si.edu/x/ view.pl?nomid=97431.

21. Tandem Web site, http://www.tandem.com/cust_ref/ dhcdsscr/dhcdsscr.htm.

22. Sun Web site, http://www.pcinews.com/business/pci//sun/ mar/gap.html.

23. Cognos Web site, http://www.cognos.com/busintell/ sutterhome.html.

24. Dallas Systems Web site, http://www.dalsys.com/integrat .htm.

25. Manugistics Web site, http://www.manugistics.com/html .tf.html.

26. McHugh Freeman Web site, http://mfa.com/pr06-30.html.

27. Smithsonian Innovation Entry, http://innovate.si.edu/x/ view.pl?nomid=97034.

28. Louis V. Gerstner (keynote speech presented at the Gartner Symposium, Orlando, Fla., October 9. 1991).

■ CHAPTER 6

1. Lewis Platt, "Creating the Extended Enterprise" (presented at the net.profit Executive Summit, Geneva, Switzerland, February 7, 1997).

2. Octel Web site, http://www.octel.com/business.customers/ cox.html.

3. Curtis F. Franklin, Jr., "Beyond Client/Server," *CIO* (1 March 1997).

4. Computer Associates Web site, http://cai.com/press/97mar/ a-1.htm.

5. Jean S. Bozman, "Jets and the Net," *Computerworld* (31 July 1995).

6. Smithsonian Innovation Entry, http://innovate.si.edu/x/ view.pl?nomid=97105.

7. "Winner Profile: MacGregor Medical Association," *CIO* (1 February 1997).

8. Alan Radding, "Season for Change," *Computerworld* (March 1997).

9. Dan Orzech, "Schwab's Bold IT Strategy Yields Wall Street Triumph," *Cambridge Technology Report,* 1996.
10. Tim W. Ferguson, "Do It Yourself," *Forbes* (22 April 1996).
11. Matthew Schifrin, "Cyber-Schwab," *Forbes* (5 May 1997).
12. Alyssa A. Lappen, "Chuck Schwab's Search for the Next Paradigm," *Institutional Investor* (April 1996).
13. Dean Tomasula, "On the Cutting Edge," *Wall Street & Technology,* vol. 14, no. 12 (December 1996).
14. Alan Ampolsk, Fleishman-Hilliard, E-mail to author, 18 July 1997.
15. Smithsonian Innovation Entry, http://innovate.si.edu/x/view.pl?nomid97046.
16. Smithsonian Innovation Entry, http://innovate.si.edu/x/view.pl?97478.
17. Informix Web site, http://www.informix.com/informix/solution/manufact/2063773/2063773.html.

■ CHAPTER 7

1. William Gates, "The Digital Nervous System" (keynote speech presented at the 1997 Microsoft CEO Summit, Seattle, Wash., May 8, 1997).
2. "SuperSite Teams with IDG's Web Publishing Inc. to Provide Job Banks for Online Publications" (SuperSite release, 15 June 1997).
3. Resumix Web site, http://www.resumix.com/products/upsa.html.
4. Resumix Web site, http://www.resumix.com/products/texasinstruments.html.
5. Resumix Web site, http://www.resumix.com/products/net.html.
6. Talisman Technology Web site, http://www.talismantech.com/ford.html.
7. Barry Hall and Kathleen Collins, "The Intramural Benefit," *Risk & Insurance* (July 1996).
8. "NetDynamics Teams with Leading Systems Integrators to Launch 'Internet in a Month' for PeopleSoft HR" (NetDynamics release, 1 May 1997).
9. Ibid.
10. SAP Web site, http://www.sap.com/usa/index.htm.

11. Ibid.
12. Hunter Group Web site, http://www.hunter-group.com/thg/art/case1.htm.
13. "Hewlett-Packard Packs Power into Personnel and Manufacturing Programs" (PeopleSoft document, June 1997).
14. "Smooth Ride for Laidlaw's HR Operations" (PeopleSoft document, November 1996).
15. Microsoft Web site, http://www.microsoft.com/industry/acc/press/dixie.htm.
16. Campbell Software Web site, http://campbellsoft.com/afitch1.htm.
17. Lee The, "HR App Meets Critical Needs," *Datamation* (June 15, 1995).
18. Smithsonian Innovation Entry, http://innovate.si.edu/x/view.pl?nomid=97442.
19. Geac Web site, http://www.smartstream.geac.com/success.

■ CHAPTER 8

1. Microsoft Web site, http://www.microsoft.com/industry/acc/casecase1329.htm.
2. SAP Web site, http://www.sap.com/success.
3. SAP Web site, http://www.sap.com/success/monier.htm.
4. SAP Web site, http://www.sap.com/success.
5. Ibid.
6. Geac Web site, http://www.smartstream.geac.com/success/bcca.htm.
7. Lou Harm, Burson-Marstellar, E-mail to author, 10 March 1997.

■ CHAPTER 9

1. Smithsonian Innovation Entry, http://innovate.si.edu/x/view.pl?nomid=97480.
2. Smithsonian Innovation Entry, http://innovate.si.edu/x/view.pl?nomid=97307.

3. PictureTel Web site, http://www.pictel.com/apps/applications/design.html.
4. Lotus Web site, http://www2.lotus.com/stories.nsf.
5. Netscape Web site, http://home.netscape.com/comprod/at_work/customer_profiles/columbia.html.
6. Avantos Web site, http://www.avantos.com/mpmune/kayser.html.
7. "Managers at Claremont Technology Group, Inc. Are Getting Candid Appraisals of Their Performance," *Computerworld* (17 March 1997).
8. VTEL Web site, http://www.vtel.com/solution/corp/cumm.html.
9. VTEL Web site, http://www.vtel.com/solution/corp/norfolk.html.
10. VTEL Web site, http://www.pictel.com/apps/applications/herm.html.
11. PictureTel Web site, http://www.pictel.com/apps/applications/designhtml.
12. Netscape Web site, http://home.netscape.com/comprod/at_work/customer_profiles/3m.html.
13. PictureTel Web site, http://www.pictel.com/apps/applications/bombardier.html.
14. North Bay (Calif.) Council Web site, http://www.nbn.com/home/nbc/telecommute.html.
15. PictureTel Web site, http://www.pictel.com/apps/applications/bostonu.html.
16. PictureTel Web site, http://www.pictel.com/apps/applications/lam.html.

■ CHAPTER 10

1. Gretchen Schaffer, Waggener-Edstrom, E-mail to author, 24 July 1997.
2. Ibid.
3. Lou Harm, Burson-Marstellar, E-mail to author, 10 March 1997.
4. Compaq Web site, http://www.compaq.com/solutions/stories/mscase.html.

Index